Monitoring and Verification of Bioremediation

BIOREMEDIATION

The *Bioremediation* series contains collections of articles derived from many of the presentations made at the First, Second, and Third International In Situ and On-Site Bioreclamation Symposia, which were held in 1991, 1993, and 1995 in San Diego, California.

First International In Situ and On-Site Bioreclamation Symposium
1(1) *On-Site Bioreclamation: Processes for Xenobiotic and Hydrocarbon Treatment*
1(2) *In Situ Bioreclamation: Applications and Investigations for Hydrocarbon and Contaminated Site Remediation*

Second International In Situ and On-Site Bioreclamation Symposium
2(1) *Bioremediation of Chlorinated and Polycyclic Aromatic Hydrocarbon Compounds*
2(2) *Hydrocarbon Bioremediation*
2(3) *Applied Biotechnology for Site Remediation*
2(4) *Emerging Technology for Bioremediation of Metals*
2(5) *Air Sparging for Site Bioremediation*

Third International In Situ and On-Site Bioreclamation Symposium
3(1) *Intrinsic Bioremediation*
3(2) *In Situ Aeration: Air Sparging, Bioventing, and Related Remediation Processes*
3(3) *Bioaugmentation for Site Remediation*
3(4) *Bioremediation of Chlorinated Solvents*
3(5) *Monitoring and Verification of Bioremediation*
3(6) *Applied Bioremediation of Petroleum Hydrocarbons*
3(7) *Bioremediation of Recalcitrant Organics*
3(8) *Microbial Processes for Bioremediation*
3(9) *Biological Unit Processes for Hazardous Waste Treatment*
3(10) *Bioremediation of Inorganics*

Bioremediation Series Cumulative Indices: 1991-1995

For information about ordering books in the Bioremediation series, contact Battelle Press. Telephone: 800-451-3543 or 614-424-6393. Fax: 614-424-3819. Internet: sheldric@battelle.org.

Monitoring and Verification of Bioremediation

Edited by

Robert E. Hinchee
Battelle Memorial Institute

Gregory S. Douglas
Arthur D. Little, Inc.

Say Kee Ong
Iowa State University

BATTELLE PRESS
Columbus • Richland

Library of Congress Cataloging-in-Publication Data

Hinchee, Robert E.
    Monitoring and verification of bioremediation / edited by Robert E.
    Hinchee, Gregory S. Douglas, Say Kee Ong.
        p.   cm.
    Includes bibliographical references and index.
    ISBN 1-57477-006-3 (hc : acid-free paper)
    1. Bioremediation—Congresses. I. Hinchee, Robert E. II. Douglas,
    Gregory S. III. Ong, Say Kee.
    TD192.5.M66 1995
    628.5′2—dc20   95-32266
                                                                     CIP

Printed in the United States of America

Additional copies may be ordered through:
Battelle Press
505 King Avenue
Columbus, Ohio 43201, USA
1-614-424-6393 or 1-800-451-3543
Fax: 1-614-424-3819
Internet: sheldric@battelle.org

CONTENTS

Foreword vii

Assessing Bioremediation of Crude Oil in Soils
and Sludges 1
S. J. McMillen, N. R. Gray, J. M. Kerr,
A. G. Requejo, T. J. McDonald, and
G. S. Douglas

Field Desorption Mass Spectroscopy Monitoring of
Changes in Hydrocarbon Type Composition During
Petroleum Biodegradation 11
M. H. Huesemann

Nutrient-Stimulated Biodegradation of Aged
Refinery Hydrocarbons in Soil 19
E. N. Drake, K. E. Stokley, P. Calcavecchio,
R. E. Bare, S. J. Rothenburger, G. S. Douglas,
and R. C. Prince

Characterization of Petroleum Biodegradation
Patterns in Weathered Contaminated Soils 29
J. T. Novak, D. Schuman, and W. Burgos

Overview of Biomolecular Methods for Monitoring
Bioremediation Performance 39
F. J. Brockman

Phospholipid Analysis of Extant Microbiota for
Monitoring In Situ Bioremediation Effectiveness 49
H. C. Pinkart, D. B. Ringelberg, J. O. Stair,
S. D. Sutton, S. M. Pfiffner, and D. C. White

Enzyme Assays as Indicators for Biodegradation 59
J. J. van der Waarde, E. J. Dijkhuis, M.J.C. Henssen,
and S. Keuning

Development of Phylogenetic Probes Specific to
Burkholderia cepacia G4 65
J. Zhou and J. M. Tiedje

Modeling Shoreline Bioremediation: Continuous Flow
and Seawater Exchange Columns 77
S. Ramstad, P. Sveum, C. Bech, and L.-G. Faksness

Physical Modeling of Shoreline Bioremediation:
Continuous Flow Mesoscale Basins 87
    P. Sveum, S. Ramstad, L.-G. Faksness, C. Bech,
    and B. Johansen

Biodegradability Study of High-Erucic-Acid-
Rapeseed-Oil-Based Lubricant Additives 97
    E. Zhou, A. Shanahan, W. Mammel, Jr., and
    R. L. Crawford

Detecting Organic Groundwater Contamination
Using Electrical Resistivity and VLF Surveys 105
    A. K. Benson, K. L. Payne, and M. A. Stubben

Continuous Bioventing Monitoring Using a
New Sensor Technology 115
    D. X. Li

Degradation Tests with PAH-Metabolizing Soil
Bacteria for In Situ Bioremediation 127
    G. Maue and W. Dott

Evaluation and Cost Assessment for Residual
Military Pollution in Germany 135
    D. Weth, W. Schröder, and J. Korr

A Bench-Scale Bioreactor to Remediate Dredged
Sludge and Soil 141
    M. Carpels, L. Kinnaer, D. Vanhoutven,
    H. Elslander, S. VanRoy, L. Hooybergs, and L. Diels

Composting Oily Sludges: Characterizing Microflora
Using Randomly Amplified Polymorphic DNA 147
    A. Persson, M. Quednau, and S. Ahrné

Hierarchy of Treatability Studies for Assured
Bioremediation Performance 157
    D. L. Saber

Determining Cleanup Levels in Bioremediation:
Quantitative Structure Activity Relationship
Techniques 165
    V.R.J. Arulgnanendran and N. Nirmalakhandan

An Application of Adaption-Innovation Theory
to Bioremediation 175
   L. J. Guerin and T. F. Guerin

A Bench-Scale Biotreatability Methodology to
Evaluate Field Bioremediation 185
   A. G. Saberiyan, J. R. MacPherson, Jr.,
   J. S. Andrilenas, R. Moore, and A. J. Pruess

Monitoring Bioremediation of Weathered Diesel
NAPL Using Oxygen Depletion Profiles 193
   G. B. Davis, C. D. Johnston, B. M. Patterson,
   C. Barber, M. Bennett, A. Sheehy, and M. Dunbavan

Protocol Development for Determining Kinetics
of In Situ Bioremediation 203
   H. H. Tabak, R. Govind, S. Pfanstiel, C. Fu,
   X. Yan, and C. Gao

Vadose Zone Nonaqueous-Phase Liquid Characterization
Using Partitioning Gas Tracers 211
   G. A. Whitley, Jr., G. A. Pope, D. C. McKinney,
   B. A. Rouse, and P. E. Mariner

Toxicity Screening of Waste Products Using
Cell Culture Techniques 223
   M. Petitmermet, A. Favre, B. Shah, U. Rösler,
   J. Mayer, and E. Wintermantel

The Biodegradation of Fluoranthene as Monitored
Using Stable Carbon Isotopes 233
   B. A. Trust, J. G. Mueller, R. B. Coffin,
   and L. A. Cifuentes

Stable Carbon Isotope Analysis to Verify
Bioremediation and Bioattenuation 241
   K. D. Van de Velde, M. C. Marley, J. Studer,
   and D. M. Wagner

Author List 259

Keyword Index 267

FOREWORD

This book and its companion volumes (see overleaf) comprise a collection of papers derived from the Third International In Situ and On-Site Bioreclamation Symposium, held in San Diego, California, in April 1995. The 375 papers that appear in these volumes are those that were accepted after peer review. The editors believe that this collection is the most comprehensive and up-to-date work available in the field of bioremediation.

Significant advances have been made in bioremediation since the First and Second Symposia were held in 1991 and 1993. Bioremediation as a whole remains a rapidly advancing field, and new technologies continue to emerge. As the industry matures, the emphasis for some technologies shifts to application and refinement of proven methods, whereas the emphasis for emerging technologies moves from the laboratory to the field. For example, many technologies that can be applied to sites contaminated with petroleum hydrocarbons are now commercially available and have been applied to thousands of sites. In contrast, there are as yet no commercial technologies commonly used to remediate most recalcitrant compounds. The articles in these volumes report on field and laboratory research conducted both to develop promising new technologies and to improve existing technologies for remediation of a wide spectrum of compounds.

The editors would like to recognize the substantial contribution of the peer reviewers who read and provided written comments to the authors of the draft articles that were considered for this volume. Thoughtful, insightful review is crucial for the production of a high-quality technical publication. The peer reviewers for this volume were:

Pradeep K. Aggarwal, *Argonne National Laboratory*
Bruce Alleman, *Battelle Columbus*
Pedro J. Alvarez, *University of Iowa*
James E. Anderson, *Stanford University*
James L. Baer, *Brigham Young University*
Serge Baghdikian (USA)
Marc Baviere, *Institut Français du Pétrole*
Donna L. Bedard, *GE Corporate R&D Center*
Kate G. Bedore, *State of Utah*
Alan D. Bettermann, *Biorenewal Technologies, Inc.*
Barbara J. Butler, *University of Waterloo* (Canada)
Patricia J. S. Colberg, *University of Wyoming*
Peter T. Cummings, *University of Tennessee*
Carol A. Daly, *Groundwater Technology, Inc.*
Jean M. Dasch, *General Motors Corporation*
Paula K. Donnelly, *Sante Fe Junior College*
Maureen A. Dooley, *ABB Environmental Services, Inc.*

Alfred Duba, *Lawrence Livermore National Laboratory*
Jean Ducreux, *Institut Français du Pétrole*
Larry A. Dudus, *Parsons Engineering Science, Inc.*
David D. Emery, *Bioremediation Service, Inc.*
Pat C. Faessler, *Bioremediation Service, Inc.*
Françoise Fayolle, *Institut Français du Pétrole*
Samuel Fogel, *Bioremediation Consulting, Inc.*
John Glaser, *U.S. Environmental Protection Agency*
Lori Graham, *Biorenewal Technologies, Inc.*
Gary M. Grey, *HydroQual, Inc.*
Christian Grön, *Technical University of Denmark*
Matthew Grossman, *Exxon Research & Engineering*
James Guckert, *Ivorydale Technical Center*
R. Gundersen, *Aquateam A/S*
Michael H. Huesemann, *Battelle Pacific Northwest*
Jeffrey Keith, *Brigham Young University*
Cheryl A. Kelley, *U.S. Environmental Protection Agency*
Niels Kroer, *Ecol. & Micro.* (Denmark)
Zoë Lees, *University of Natal* (Rep. of South Africa)
S. Lorentz, *University of Natal* (Rep. of South Africa)
Donn Marrin, *InterPhase Environmental, Inc.*
Sara J. McMillen, *Exxon Production Research Co.*
Soren Moller, *Technical University of Denmark*
James G. Mueller, *SBP Technologies, Inc.*
Richard Ornstein, *Battelle Pacific Northwest*
Jerome Perry, *North Carolina State University*
Roger Prince, *Exxon Research & Engineering*
Charles Michael Reynolds, *U.S. Army*
Greg Smith, *Great Lakes Environmental Center*
Gregory Smith, *ENSR Consulting & Engineering*
Dirk Springael, *VITO* (Belgium)
James Staley, *University of Washington*
Beth A. Trust, *National Research Council*
F. Michael von Fahnestock, *Battelle Columbus*
John T. Wilson, *U.S. Environmental Protection Agency*
Darla Workman, *Oregon State University*
Lin Wu, *University of California, Davis*

The figure that appears on the cover of this volume was adapted from the article by McMillen et al. (see page 6).

Finally, I want to recognize the key members of the production staff, who put forth significant effort in assembling this book and its companion volumes. Carol Young, the Symposium Administrator, was responsible for the administrative effort necessary to produce the ten volumes. She was assisted by Gina Melaragno, who tracked draft manuscripts through the review process and generated much of the correspondence with the authors, co-editors, and peer reviewers. Lynn Copley-Graves oversaw text editing and directed the layout of the book, compilation of the keyword indices, and production of the camera-ready copy. She was assisted by technical editors Bea Weaver and Ann Elliot. Loretta Bahn was responsible for text processing and worked many long hours incorporating editors' revisions, laying out the camera-ready pages and figures, and maintaining the keyword list. She was assisted by Sherry Galford and Cleta Richey; additional support was provided by Susan Vianna and her staff at Fishergate, Inc. Darlene Whyte and Mike Steve proofread the final copy. Judy Ward, Gina Melaragno, Bonnie Snodgrass, and Carol Young carried out final production tasks. Karl Nehring, who served as Symposium Administrator in 1991 and 1993, provided valuable insight and advice.

The symposium was sponsored by Battelle Memorial Institute with support from many organizations. The following organizations cosponsored or otherwise supported the Third Symposium.

Ajou University–College of Engineering (Korea)
American Petroleum Institute
Asian Institute of Technology (Thailand)
Biotreatment News
Castalia
ENEA (Italy)
Environment Canada
Environmental Protection
Gas Research Institute
Groundwater Technology, Inc.
Institut Français du Pétrole
Mitsubishi Corporation
OHM Remediation Services Corporation
Parsons Engineering Science, Inc.
RIVM–National Institute of Public Health and the Environment
    (The Netherlands)
The Japan Research Institute, Limited
Umweltbundesamt (Germany)
U.S. Air Force Armstrong Laboratory–Environics Directorate
U.S. Air Force Center for Environmental Excellence
U.S. Department of Energy Office of Technology Development
    (OTD)
U.S. Environmental Protection Agency
U.S. Naval Facilities Engineering Services Center
Western Region Hazardous Substance Research Center—Stanford
    and Oregon State Universities

Neither Battelle nor the cosponsoring or supporting organizations reviewed this book, and their support for the Symposium should not be construed as an endorsement of the book's content. I conducted the final review and selection of all papers published in this volume, making use of the essential input provided by the peer reviewers and other editors. I take responsibility for any errors or omissions in the final publication.

Rob Hinchee
June 1995

Assessing Bioremediation of Crude Oil in Soils and Sludges

Sara J. McMillen, Nancy R. Gray, Jill M. Kerr,
Adolpho G. Requejo, Thomas J. McDonald,
and Gregory S. Douglas

ABSTRACT

Standard bulk property analytical methods currently being employed to evaluate crude oil bioremediation efficacy in soils [U.S. Environmental Protection Agency (EPA) Methods 418.1 and 413.1] provide no information concerning the mechanisms by which hydrocarbon losses are occurring (e.g., biodegradation versus leaching). Site/sample heterogeneity in field bioremediation projects may make it difficult to accurately quantify hydrocarbon losses due to biodegradation. To better understand the mechanisms by which losses are occurring and to accurately evaluate biodegradation rates, the hydrocarbon analytical methods must provide both quantitative and compositional information. In this study, laboratory bioremediation experiments were used to compare the results of bulk property analytical methods with those methods used by petroleum geochemists that provide both quantitative and compositional data. A tecator extraction was used to isolate the total extractable matter (TEM) from the samples. Compositional changes were monitored by (1) column chromatography to determine class distributions, (2) high resolution gas chromatography with a flame-ionization detector (GC/FID) and (3) gas chromatography/mass spectrometry (GC/MS). Illustrations of the compositional changes detected by each method and their application to validating bioremediation are provided.

INTRODUCTION

Several analytical methods that assess changes in hydrocarbon composition can be used to quantitatively assess hydrocarbon losses due to biodegradation. Compositional changes of crude oils resulting from biodegradation have been well documented by petroleum geochemists since some crude oil deposits have been biodegraded in the subsurface oil reservoir (Hunt 1979; Connan 1984; Miles 1989). Several authors have used similar techniques to illustrate that decreases in total petroleum hydrocarbons (TPH) resulted from bioremediation processes. Gas chromatography has been used to quantify changes in *n*-alkanes and isoprenoid ratios (Oudot et al. 1987; McMillen et al. 1993) and GC/MS has been used to quantify polyaromatic hydrocarbon (PAH) losses (Elmendorf 1993; Douglas et al. 1994). Changes in class composition have been observed in biodegraded crude oils and fuel products (Jobson et al. 1972; Bailey et al. 1973; Westlake et al. 1975; Song et al. 1990). The biomarker C30-17α, 21β hopane was used to quantify oil depletion for heterogeneous sites in Prince William Sound (Prince et al. 1994) and has also been used for quantifying biodegradation losses in soil (McMillen et al. 1993). However, too frequently, field and laboratory data are reported that show only TPH losses without confirming compositional changes to illustrate that biodegradation is the mechanism of loss.

During field bioremediation projects, the observed losses in bulk total hydrocarbons may not be the result of biodegradation processes, but rather are due to dilution, leaching, or volatilization. Dilution may occur when heterogeneous oily soils are tilled so that clumps of soil with higher oil content are mixed into less oily soil. Leaching may occur during irrigation or heavy rainfall. Volatilization of lighter hydrocarbons may occur during tilling or during warm summer months. Accurately quantifying biodegradation losses in the field are also often complicated by the spatial variability of hydrocarbons at spill sites. However, by using the analytical methods described in this paper, the compositional changes in oils that result from biodegradation can be observed. One of the most useful compositional changes is that of the biomarkers or hydrocarbons which are very resistant to biodegradation. Since biomarkers such as steranes and triterpanes are resistant to biodegradation, as the amount of biodegradation increases, they will concentrate in the oil extract. Therefore, the increase in their concentration is proportional to the amount of oil degraded. Calculating the percent loss of hydrocarbons based upon biomarker concentrations eliminates the problems associated with spatial variability and abiotic losses from dilution, volatilization, or leaching.

EXPERIMENTAL PROCEDURES

Over 50 laboratory bioremediation experiments have been conducted using oily sludges, oil-spiked soils, and site soils collected from aged hydrocarbon spill areas. The sludges typically were composted by adding bulking agents such as manure and wood chips. The soils were treated in 2- to 4-kg mesocosms. Nitrogen and phosphorus fertilizers were added to enhance the rate of biodegradation. Total hydrocarbon concentration and compositional changes were observed over time for all experiments. The TEM concentrations were obtained by a tecator extraction with methylene chloride:methanol (9:1) as the solvent. The extract was dried at 40°C with a nitrogen stream to a constant weight. This method yields the C12 to C15+ boiling point fraction. Class distributions were characterized by precipitation (asphaltenes) and alumina-silica gel chromatography (saturates, aromatics, and polars). (High-pressure liquid chromatography or Iatroscan also can be used for this purpose.) GC/FID was used to obtain saturate and aromatic fraction "fingerprints" and to quantify both resolved and unresolved components of these fractions. The total resolved and unresolved hydrocarbons quantified by GC/FID are herein reported as TPH. The fractions were also analyzed by GC/MS to determine concentrations of specific polyaromatic hydrocarbons, including their alkylated homologs, and sterane and triterpane biomarkers such as C30-17α, 21β hopane (Douglas et al. 1994).

RESULTS

The relative distribution of saturates, aromatics, and polar compounds change as biodegradation progresses. Biodegradation produces a progressive decrease in the relative amounts of saturates and an increase in the aromatics and polars as shown in Figure 1a. This results from the relative biodegradability of these fractions in the order saturates > aromatics > polars. The greater biodegradability of saturates is also shown in Figure 1b by the decrease in their absolute concentrations relative to the aromatics and polars.

GC/FID provides further compositional detail on the biodegradation of hydrocarbons. Figure 2a illustrates the GC/FID traces of two oil-spiked soil samples; one an untreated control and the other a fertilizer-treated sample that has begun to biodegrade. The *n*-alkanes are greatly reduced in the biodegraded sample after 8 weeks. These hydrocarbons are the most readily biodegradable of the C12 to C15+ fraction of crude oils. Crude oils enriched in *n*-alkanes tend to show rapid TPH losses during bioremediation.

Because *n*-alkanes are more readily biodegraded than the branched-chain isoprenoids, pristane (Pr) and phytane (Ph), the ratios of C17/Pr and C18/Ph will decrease over time. After a few weeks, however, the isoprenoids will begin to biodegrade, and the ratio may then increase slightly. Eventually both the *n*-alkanes and isoprenoid concentrations will be reduced to nondetectable levels. These trends are illustrated in Figure 2b. At this stage of degradation, GC/FID traces are generally dominated by a "hump" or unresolved complex mixture (UCM). This feature represents a complex mixture of hydrocarbons that are more difficult to biodegrade.

GC/FID can also be used to monitor the fate of the UCM. Figure 3 illustrates losses for the resolved and UCM portions of the saturate and aromatic fractions. The resolved components in the saturate fraction, consisting primarily of *n*-alkanes, degrade more quickly than the saturate UCM. The aromatic UCM tends to biodegrade more slowly than the saturate UCM.

GC/MS can be used to examine the fate of compounds such as PAHs. Figure 4 illustrates the changes in concentrations of selected two to five-ring PAHs and their alkylated homologs as measured by GC/MS. Two, three, and some four-ring PAHs, including the homologs, have been found to be biodegradable. Biodegradation typically removes the parent compounds and lower alkylated homologs before affecting the higher homologs.

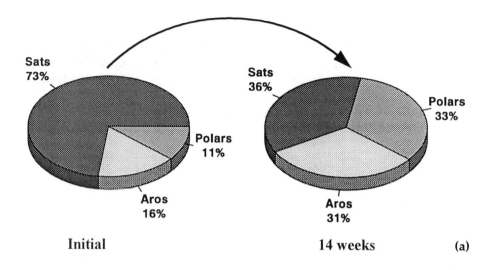

Initial   14 weeks   (a)

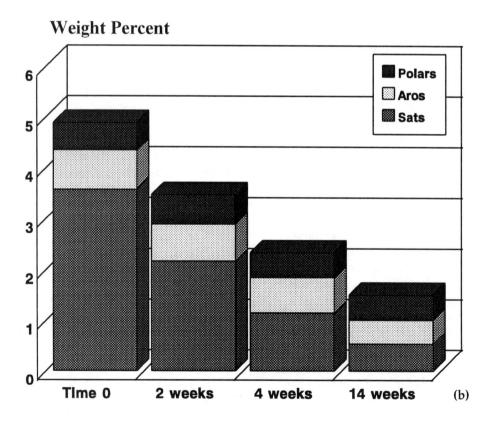

FIGURE 1. (a) Relative distribution of saturates (sats), aromatics (aros), and polars. (b) Concentration of classes on a dry soil weight basis.

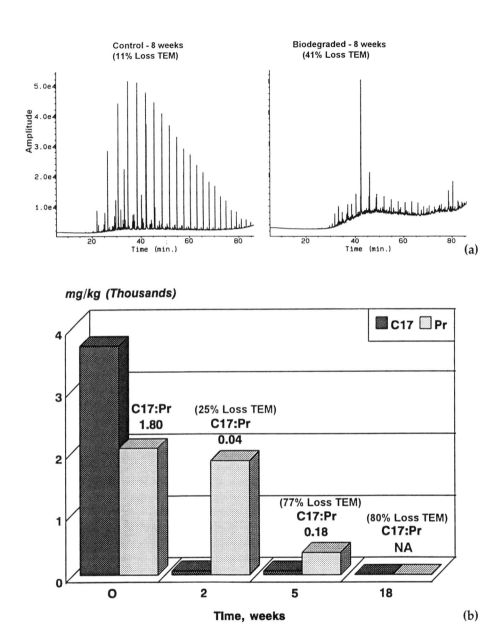

FIGURE 2. (a) GC/FID "fingerprints" of a nonbiodegraded and biodegraded
sample. (b) C17/Pr ratios versus time.

GC/MS can also be used to quantify the concentrations of polycyclic saturate
compounds such as steranes and triterpanes that are useful refractory indicators
of bioremediation in oil. As the extent of biodegradation increases, the steranes
and triterpane biomarkers will concentrate in the oil extract. The increase in

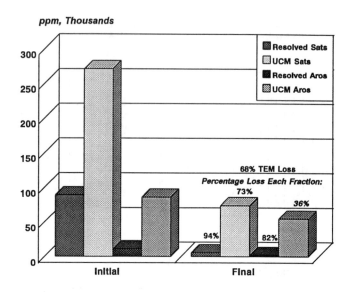

FIGURE 3. Biodegradation losses of resolved and UCM portions of the
saturates and aromatics.

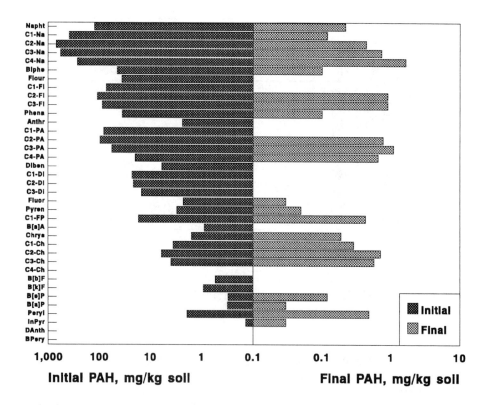

FIGURE 4. Changes in PAH concentrations resulting from biodegradation.

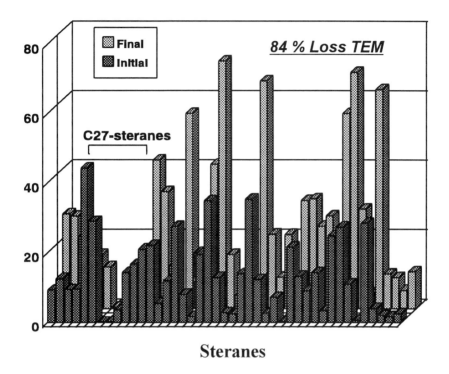

Steranes

FIGURE 5. Biodegradation of C27 steranes in heavily biodegraded sample (concentration on oil extract basis, ppm).

their concentration is proportional to the amount of oil degraded. Figure 5 illustrates depletion of the C27 steranes in the advanced stages of bioremediation, while other steranes have increased in concentration. The triterpane C30-17α, 21β hopane, which is frequently the most abundant triterpane in crude oils, has been particularly useful for quantifying total oil depletion. Figure 6 illustrates the results for total oil depletion determined from the increase in hopane concentrations in comparison to losses of TPH measured by GC/FID.

DISCUSSION

Hydrocarbon losses resulting from biodegradation are reflected in the compositional changes in the remaining petroleum fractions. Compositional changes occur because of differences in the relative biodegradability of specific classes of hydrocarbons. GC/FID is a particularly useful tool to observe biodegradation since it allows for quantification of hydrocarbons and visually illustrates changes in composition resulting from the preferential biodegradation of the *n*-alkanes. Ratios of C17 and C18 to isoprenoids are useful for fresh crude oil spills in the early stages of bioremediation (a few weeks to months). However, these ratios

Hopane, % Depletion

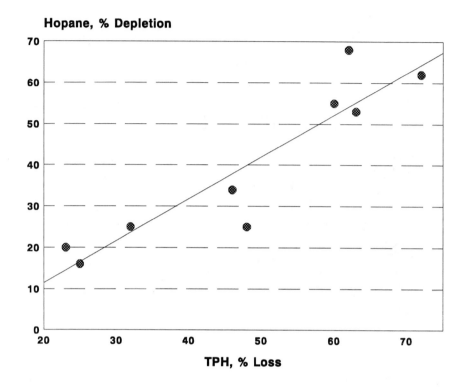

TPH, % Loss

FIGURE 6. Comparison of hopane depletion results with TPH (by GC/FID).

are less useful for old hydrocarbon spills or advanced bioremediation projects that are highly weathered and biodegraded. For these soils, less biodegradable compounds such as triterpanes must be used. The concentration of triterpanes in the residual oil will increase as the extent of biodegradation increases. The ratio of C30-17α, 21β hopane in the initial and biodegraded extracts can be used to quantify oil depletion. The use of these analytical methods is recommended to provide information on the mechanism of hydrocarbon losses and to better quantify biodegradation rates.

REFERENCES

Bailey, N.J.L., A. M. Jobson, and M. A. Rogers. 1973. "Bacterial Alteration of Crude Oil: Comparison of Field and Experimental Data," *Chemical Geology 11*: 203-221.

Connan, J. 1984. "Biodegradation of Crude Oils in Reservoirs." In J. Brooks and D. Welte (Eds.), *Advances in Petroleum Geochemistry*, Vol. 1, pp. 299-335. Academic Press, London.

Douglas, G. S., R. C. Prince, E. L. Butler, and W. G. Steinhauer. 1994. "The Use of Internal Chemical Indicators in Petroleum and Refined Products to Evaluate the Extent of Biodegradation." In R. E. Hinchee, B. C. Alleman, R. E. Hoeppel, and R. N. Miller (Eds.), *Hydrocarbon Bioremediation*, pp. 219-235. Lewis Publishers, Ann Arbor, MI.

Elmendorf, D. L., C. E. Haith, G. S. Douglas, and R. C. Prince. 1994. "Relative Rates of Biodegradation of Substituted Polyaromatic Hydrocarbons." In R. E. Hinchee, A. E. Leeson, L. Semprini, and S. K. Ong (Eds.), *Bioremediation of Chlorinated and Polycyclic Aromatic Hydrocarbons.* Lewis Publishers, Ann Arbor, MI.

Hunt, J. M. 1979. *Petroleum Geochemistry and Geology,* pp. 382-387. W. H. Freeman, San Francisco, CA.

Jobson, A., F. D. Cook, and D. W. S. Westlake. 1972. "Microbial Utilization of Crude Oil." *Applied Microbiology*, 23(6): 1082-1089.

McMillen, S. J., J. M. Kerr, and N. R. Gray. 1993. "Microcosm Studies of Factors That Influence Bioremediation of Crude Oils in Soil." SPE 25981, presented at the Society for Petroleum Engineers/Environmental Protection Agency Exploration and Production Environmental Conference, San Antonio, TX.

Miles, J. A. 1989. *Illustrated Glossary of Petroleum Geochemistry,* pp. 31-32. Oxford University Press, New York, NY.

Prince, R. C., D. L. Elmendorf, J. R. Lute, C. S. Hsu, C. E. Haith, J. D. Senius, G. J. Dechert, G. S. Douglas, and E. L. Butler. 1994. "17α(H), 21β(H)-Hopane as a Conserved Internal Marker for Estimating the Biodegradation of Crude Oil." *Environmental Science and Technology*, 28: 142-145.

Oudot, J., P. Fusey, D. E. Abdelouahid, S. Haloui, and M. F. Roquebert. 1987. "Degrading Capacities of Bacteria and Fungi Isolated from a Soil Contaminated by a Fuel." *Canadian Journal of Microbiology*, 33: 232-243.

Song, H., X. Wang, and R. Bartha. 1990. "Bioremediation Potential of Terrestrial Fuel Spills." *Applied and Environmental Microbiology*, 56: 652-656.

Westlake, D.W.S., W. Belicek, A. Jobson, and F. D. Cook. 1975. "Microbial Utilization of Raw and Hydrogenated Shale Oils." *Canadian Journal of Microbiology*, 22: 221-227.

Field Desorption Mass Spectroscopy Monitoring of Changes in Hydrocarbon Type Composition During Petroleum Biodegradation

Michael H. Huesemann

ABSTRACT

A comprehensive petroleum hydrocarbon characterization procedure involving group type separation, boiling point distribution, and hydrocarbon typing by field desorption mass spectroscopy (FDMS) has been developed to quantify changes in hydrocarbon type composition during bioremediation of petroleum-contaminated soils. FDMS is able to quantify the concentration of hundreds of specific hydrocarbon types based on their respective hydrogen deficiency (z-number) and molecular weight (carbon number). Analytical results from two bioremediation experiments involving soil contaminated with crude oil and motor oil indicate that alkanes and two-ring saturates (naphthenes) were readily biodegradable. In addition, low-molecular-weight hydrocarbons generally were biodegraded to a larger extent than those of high molecular weight. More importantly, it was found that the extent of biodegradation of specific hydrocarbon types was comparable between treatments and appeared to be unaffected by the petroleum contaminant source, soil type, or experimental conditions. It was therefore concluded that in these studies the extent of total petroleum hydrocarbon (TPH) biodegradation is primarily affected by the molecular composition of the petroleum hydrocarbons present in the contaminated soil.

INTRODUCTION

Numerous studies have been performed to determine the fate of either individual hydrocarbon compounds such as polycyclic aromatics (Bossert & Bartha 1986; Huesemann & Moore 1993; Sims & Overcash 1983), *n*-alkanes (Kennicutt 1988; Walker & Colwell 1973), and biomarkers (Butler et al. 1991; McMillen et al. 1993) or entire hydrocarbon classes such as saturates and aromatics (McMillen et al. 1993; Huesemann & Moore 1993) during bioremediation of petroleum-contaminated soils. This paper introduces a new analytical hydro-carbon characterization procedure which can be used to monitor changes in the concentration of several hundred hydrocarbon types present in TPH during bioremediation treatment.

MATERIAL AND METHODS

Bioremediation Treatments

Treatment A. Highly weathered crude oil-contaminated site soil was sieved (Tyler screen #4) and amended with cow manure, refinery-activated sludge solids (inoculum), ammonium nitrate, and potassium phosphate (Huesemann & Moore 1993). After pH and moisture adjustment, the contaminated soil was land-treated for a 52-week period in a closed landfarming mesocosm (Huesemann 1994a).

Treatment B. 10 kg of a sandy loam was spiked with approximately 1 kg of fresh (unused) 10W-30 motor oil. After weathering and fertilizer amendments, the soil was slurry-treated for 52 weeks (Huesemann 1994b). Samples for compositional analyses were taken initially and after 34 weeks of slurry bioremediation.

Comprehensive Hydrocarbon Characterization Scheme

Initial and final samples from bioremediation treatments were subjected to an extensive petroleum hydrocarbon characterization strategy (Figure 1). The soil or soil slurry sample was soxhlet Freon™ extracted and a fraction of the resulting oil and grease (O&G) extract was dried to obtain the gravimetric O&G concentration. The remainder of the O&G extract was treated with silica gel to remove polar compounds. After evaporation of the Freon™ under nitrogen at room temperature, the hydrocarbon concentration in the resulting TPH extract was measured gravimetrically (TPH-GR). A portion of the gravimetric TPH extract was redissolved in cyclopentane and separated into a saturate and aromatic fraction by column chromatography. Both saturate and aromatic fractions were further characterized by a boiling point profile (BPP) analysis and field desorption mass spectroscopy (FDMS). The BPP gives detailed information regarding the molecular weight distribution from C_{13} to C_{44} as well as the fraction of any hydrocarbons heavier than C_{44}. The FDMS analysis provides compositional data for any C_{13}-C_{44} hydrocarbons based on both carbon number (or molecular weight) and hydrogen deficiency (or z-number). For details, refer to Huesemann (1995a).

Data Analysis and Interpretation

The power of the comprehensive hydrocarbon characterization scheme (Figure 1) is related to the fact that all extracted hydrocarbons with carbon numbers greater than 13 can be accounted for on a mass balance basis. Data from the group type separation (GTS) analysis in conjunction with the BPPs are used to

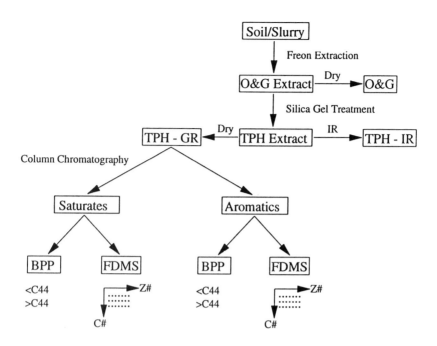

FIGURE 1. Comprehensive analytical characterization procedure for petroleum hydrocarbons in contaminated soils or slurries (adopted from Huesemann, 1995a).

differentiate the TPH compounds into four broad classes, namely $<C_{44}$-saturates, $>C_{44}$-saturates, $<C_{44}$-aromatics, and $>C_{44}$-aromatics. FDMS of the saturate and aromatic fractions provides more detail regarding the composition of $<C_{44}$ compounds in each fraction because mass spectroscopy signals are integrated only in the C_{13} to C_{44} range.

The field desorption mass spectrum provides semiquantitative data regarding the concentration of hydrocarbon types in terms of their respective carbon number (n) and hydrogen deficiency (z-number) according to the general molecular formula C_nH_{2n+z} (Boduszynski, 1987; Dzidic et al., 1992). The z-number is used to infer the approximate molecular ring structure of a hydrocarbon. For example, a saturate hydrocarbon with a z-number of −4 has three saturate rings, while an aromatic hydrocarbon with a z-number of −6 is characterized by one aromatic ring. Examples of hydrocarbon structures based on z numbers are presented in Figure 2. Since FDMS also provides concentration data regarding the carbon number of hydrocarbon types, it is possible to estimate the approximate length of side chains (R) attached to the ring structures. For instance, FDMS data may indicate that the $<C_{44}$ saturate fraction contains a hydrocarbon type with C = 20 and z = −2 at a concentration of 5% (wt). Thus, 5% (wt) of the hydrocarbons in the saturate fraction consists of two-ring saturates (Decalin, C = 10) with one or several side chains (methyl, ethyl, propyl-groups, etc.) containing a total of

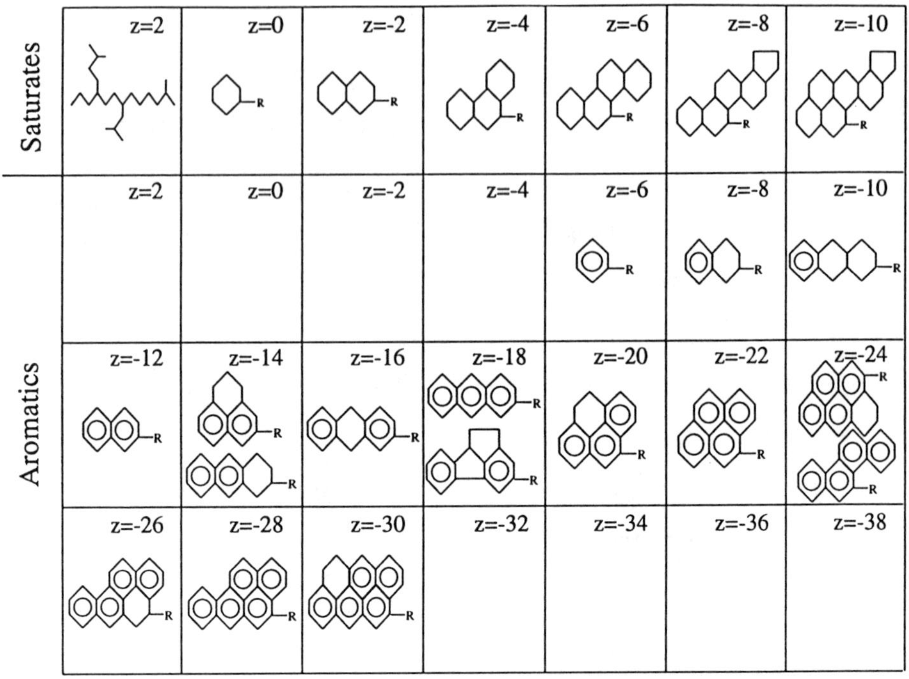

	z=2	z=0	z=-2	z=-4	z=-6	z=-8	z=-10
Saturates							

	z=2	z=0	z=-2	z=-4	z=-6	z=-8	z=-10

	z=-12	z=-14	z=-16	z=-18	z=-20	z=-22	z=-24
Aromatics							
	z=-26	z=-28	z=-30	z=-32	z=-34	z=-36	z=-38

FIGURE 2. Example molecular structures for saturate and aromatic hydrocarbons as a function of z-number and molecular weight (adopted from
Huesemann, 1995a). Note that FDMS cannot differentiate between structures characterized by z-numbers 14 units apart. Consequently, several
aromatic rings structures exist depending upon the molecular weight of the
hydrocarbon. The "R" group symbolizes any possible number of alkyl side
groups with variable carbon chain lengths. No aromatic structures exist
for z > −4 and no additional molecular structures are shown for z < −32.

10 carbon atoms in their backbones. Finally, because FDMS cannot differentiate
between z-numbers 14 units (i.e., protons) apart, several different ring structures
are possible depending on the carbon number of the respective hydrocarbon
type (Figure 2). For instance, an aromatic hydrocarbon of carbon number 18
and z-number −8 or −22 could be either a two-ring (1 aromatic and 1 saturate
ring) molecule with a C_8 side chain or a four-ring aromatic with a C_2 side chain.
For a numerical example to calculate the concentration of specific hydrocarbon
types, see Huesemann (1995b).

RESULTS AND DISCUSSION

Since it would be impossible to display initial and final concentrations of
all 434 hydrocarbon types [14 (2 × 7) z-types for a carbon number range of C_{13} to

C_{44}] in each treatment, only compositional changes of selected hydrocarbon types (defined by z-number and carbon number) during bioremediation will be presented in Figures 3, 4 and 5. Figure 3 shows the initial and final concentrations of straight-chain and branched-chain alkanes (z = 2 saturates) as a function of carbon number during bioremediation in treatment A. It is evident that, independent of the molecular weight or carbon number (i.e., length) of the respective hydrocarbon type, all z = 2 saturated compounds are almost completely biodegraded. Figures 4 and 5 present compositional changes in dicyclic naphthenes (z = −2 saturates) during bioremediation in treatments A and B, respectively. Comparison of these two concentration profiles indicates that the initial composition of dicyclic saturates differs substantially between the crude oil of treatment A and the motor oil of treatment B. For example, the highest concentration of dicyclic saturates in the motor oil were found in the carbon number range of approximately 25 to 30 while the initial concentration profile for dicyclic hydrocarbons was skewed toward lower carbon numbers (C < 25) in the crude oil-contaminated soil. In addition, there is a substantial difference in concentrations of individual hydrocarbon types between treatments. For example, the highest initial concentration of any dicyclic hydrocarbon type was only 229 mg/kg (C = 20) in treatment A, while the concentration of dicyclic saturates in treatment B was as high

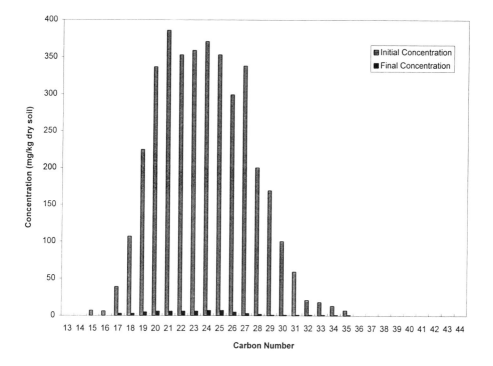

FIGURE 3. Initial and final concentrations of z = 2 saturates (straight and branched-chain alkanes) as a function of carbon number during bioremediation in treatment A.

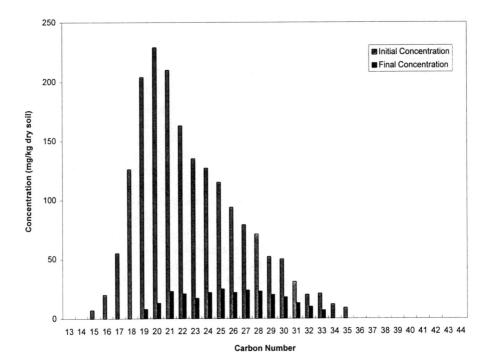

FIGURE 4.   Initial and final concentrations of z = –2 saturates (dicyclic
naphthenes) as a function of carbon number during bioremediation in
treatment A.

as 1,412 mg/kg (C = 27). Despite differences in the carbon number distribution
and initial concentrations of dicyclic hydrocarbon types between treatments A
and B, it appears that in both treatments low-molecular-weight hydrocarbons
were completely biodegraded while the extent of biodegradation of dicyclics
characterized by higher carbon numbers was on average approximately 75%.
These observations indicate that the extent of hydrocarbon type biodegradation
is affected primarily by the molecular structure (i.e., the number of saturate rings
and the carbon number) and that differences in treatment conditions (land treatment vs. slurry treatment, fertilizer concentrations, soil type) and contaminant
source (crude oil vs. motor oil) do not appear to significantly influence hydrocarbon type biodegradation in these two soil bioremediation studies (Huesemann
1995a & 1995b).

REFERENCES

Boduszynski, M. M. 1987. "Composition of Heavy Petroleums. 1. Molecular Weight,
    Hydrogen Deficiency, and Heteroatom Concentration as a Function of Atmospheric
    Equivalent Boiling Point up to 1400°F (760°C)." *Energy and Fuels* 1: 2-11.

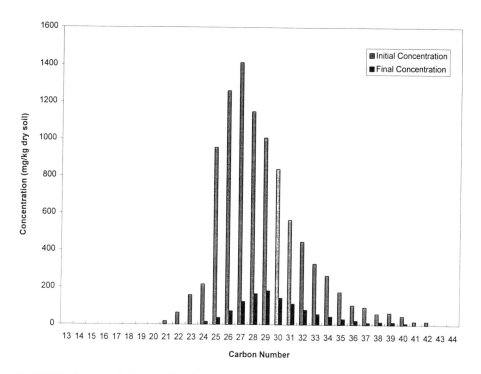

FIGURE 5. Initial and final concentrations of z = −2 saturates (dicyclic naphthenes) as a function of carbon number during bioremediation in treatment B.

Bossert, I. D., and R. Bartha. 1986. "Structure Biodegradability Relationships for Polycyclic Aromatic Hydrocarbons in Soil." *Bull. Environ. Contam. Toxicol. 37*:490-495.

Butler, E. L., G. S. Douglas, W. G. Steinhauer, R. C. Prince, T. Aczel, C. S. Hsu, M. T. Bronson, J. R. Clark, and J. E. Lindstrom. 1991. "Hopane, a New Chemical Tool for Measuring Oil Biodegradation." In: R. E. Hinchee and R. F. Olfenbuttel (Eds.), *On-Site Bioreclamation: Processes for Xenobiotic and Hydrocarbon Treatment.* Butterworth-Heinemann, Stoneham, MA. pp. 515-521.

Dzidic, I., H. A. Petersen, P. A. Wadsworth, and H. V. Hart. 1992. "Townsend Discharge Nitric Oxide Chemical Ionization Gas Chromatography/Mass Spectrometry for Hydrocarbon Analysis of the Middle Distillates." *Analytical Chemistry, 64:* 2227-2232.

Huesemann, M. H. 1994a. "Guidelines for Landtreating Petroleum Hydrocarbon Contaminated Soils." *Journal of Soil Contamination, 3:*299-318.

Huesemann, M. H. 1994b. "Comparison between Solid-phase and Slurry-phase Bioremediation of Diesel and Motor Oil Contaminated Soils." *Proceedings of the IGT Symposium on Gas, Oil, and Environmental Biotechnology.* Colorado Springs, December 1994. Submitted.

Huesemann, M. H. 1995a. "Predictive Model for Estimating the Extent of Petroleum Hydrocarbon Biodegradation in Contaminated Soils." *Environmental Science and Technology 29*(1):7-18.

Huesemann, M. H. 1995b. "Changes in Hydrocarbon Type Composition during Bioremediation of Petroleum Hydrocarbon Contaminated Soils." *Environmental Toxicology and Chemistry.* Submitted for publication.

Huesemann, M. H., and K. O. Moore. 1993. "Compositional Changes during Landfarming of Weathered Michigan Crude Oil-Contaminated Soil." *Journal of Soil Contamination 2*:245-264.

Kennicutt, M. C. 1988. "The Effect of Biodegradation on Crude Oil Bulk and Molecular Composition." *Oil and Chemical Pollution.* 4:89-112.

McMillen, S. J., J. M. Kerr, and N. R. Gray. 1993. "Microcosm Studies of Factors that Influence Bioremediation of Crude Oils in Soil." *Proceedings of SPE/EPA Exploration and Production Environmental Conference.* San Antonio, Texas. pp. 389-400.

Sims, R. C., and M. R. Overcash. 1983. "Fate of Polynuclear Aromatic Compounds in Soil-Plant Systems." *Residue Reviews 88*:2-68.

Walker, J. D., and R. R. Colwell. 1973. "Microbial Petroleum Degradation: Use of Mixed Hydrocarbon Substrates." *Applied Microbiology, 27*:1053-1060.

Nutrient-Stimulated Biodegradation of Aged Refinery Hydrocarbons in Soil

Evelyn N. Drake, Karen E. Stokley,
Peter Calcavecchio, Richard E. Bare,
Stephen J. Rothenburger, Gregory S. Douglas,
and Roger C. Prince

ABSTRACT

Aged hydrocarbon-contaminated refinery soil was amended with water and nutrients and tilled weekly for 1 year to stimulate biodegradation. Gas chromatography/mass spectrometry (GC/MS) analysis of polycyclic aromatic compounds (PAHs) and triterpane biomarkers, and Freon IR analysis of total petroleum hydrocarbons (TPH), were used to determine the extent of biodegradation. There was significant degradation of extractable hydrocarbon (up to 60%), but neither hopane, oleanane, nor the amount of polars decreased during this period of bioremediation, allowing them to be used as conserved internal markers for estimating biodegradation. Significant degradation of the more alkylated two- and three-ring compounds, and of the four-ring species pyrene and chrysene and their alkylated congeners, was seen. Substantial degradation (>40%) of benzo(b)fluoranthene, benzo(k)fluoranthene, and benzo(a)pyrene also was seen. Our results show that bioremediation can be a useful treatment in the cleanup of contaminated refinery sites.

INTRODUCTION

The chemistry and microbiology of petroleum biodegradation in freshly contaminated soils and water has been extensively studied (Leahy and Colwell 1990; Atlas and Bartha 1992; Prince 1993). Much less has been reported on the biodegradation of aged hydrocarbons that have been subjected to natural weathering and biodegradation for decades. Our interest in in situ bioremediation as a potentially cost-effective cleanup technology for refinery soils makes it important to understand the chemistry and microbiology of biodegradation of aged hydrocarbons.

Adsorption of PAHs on soil is poorly understood, but it has been shown to influence the biodegradability and biotoxicity of the hydrocarbons (Weissenfels et al. 1992; Landrum et al. 1992). Because the adsorption occurs at very slow rates, realistic experimental samples cannot be created by adding weathered oils to soil. Sieved, well-mixed field soil therefore was used for this study.

Site heterogeneity is a major problem in evaluation of bioremediation in the field. The triterpane hopane (a bacterial fossil) has been successfully used as a conserved internal marker for estimating bioremediation of crude oil in marine spills (Prince et al. 1994; Bragg et al. 1994). It is of interest to know whether hopane or other biomarkers are conserved in well-mixed field soils subject to extensive biodegradation. One way to assess this is to compare the levels of hopane with those of other compounds present in the soil.

EXPERIMENTAL PROCEDURES

A sample of refinery soil was obtained from a potential remediation site from a depth of 0 to 2 ft (0 to 0.6 m) with a backhoe. The soil was sieved through a ½-in. (1.3-cm) screen and homogenized with a portable split-vane cement mixer in the field. The soil was further sieved in the laboratory to ¼ in. (0.6 cm) for greater sample uniformity. Soil particle size, weight percent water, water-holding capacity, pH, and nitrogen and phosphorous species were determined by standard methods (Klute 1986). Estimates of the numbers of heterotrophic and oil-degrading bacteria were determined with the most-probable-number (MPN) technique (Brown and Braddock 1990).

PAHs, their alkylated homologs, and triterpane biomarkers were determined by GC/MS of methylene chloride extracts of the soil (Douglas et al. 1992). TPH was determined by a modified U.S. Environmental Protection Agency (U.S. EPA) soxhlet extraction method 418.1 (U.S. EPA 1983).

Soil was dried with anhydrous sodium sulfate and extracted for 4 hours with Freon 113. The extract was treated with silica gel and analyzed for TPH at $2,930$ cm^{-1} with an infrared spectrophotometer. TPH results were found to be sensitive to soil moisture especially at higher levels (>15 wt %), and it was necessary to use sufficient sodium sulfate to completely dry the soil sample (9 parts by weight in this case) and to grind the dried product into a free-flowing powder prior to extraction.

Bioremediation studies were conducted in covered laboratory microcosms containing 2 kg soil at a depth of 2 to 3 in. (5 to 8 cm). The microcosms were amended with urea and disodium phosphate nutrients at three levels, and were incubated at room temperature for 1 year. The microcosms were tilled weekly and soil moisture was maintained at 50% of holding capacity. Laboratory air was passed over the surface of the soil at 25 mL/min. Composite samples were taken over time. The results reported in this paper are for 52-week amended and unamended samples. Other results will be reported as they become available.

RESULTS

Soil characteristics are listed in Table 1. The soil is a loam (based on particle size). Soluble forms of nitrogen are very low making it a good candidate for nutrient-stimulated bioremediation.

Relative concentrations of TPH, methylene chloride extractable material, hopane, oleanane, polars, and C3-phenanthrene in amended relative to unamended 52-week samples are given in Figure 1. Our nutrient amendments caused TPH to decrease as much as 60% and methylene chloride extractable material to decrease by 40%. (Methylene chloride is a more aggressive solvent than Freon, so it extracts all the petroleum hydrocarbon, but also extracts organic matter, such as humic acids, from the soil, so may overestimate the amount of petroleum hydrocarbon in a soil or sediment sample.) Hopane, oleanane (an angiosperm fossil, Peters and Moldowan 1993), and the polar material in methylene chloride extracts did not show significant degradation. Hopane was more abundant than oleanane (6- to 7-fold), so it was used to normalize PAH data (see Douglas et al. 1994; Bragg et al. 1994). Data for one of the more highly degraded PAHs, C3-phenanthrene, show that depletion of individual species can be very high. Figures 2 through 5 show the depletion of parent and methylated derivatives of naphthalene, phenanthrene, pyrene, and chrysene.

DISCUSSION

The depletion patterns of PAHs shown in Figures 2 through 5 differ from those seen in studies of fresh oil biodegradation in water (e.g., Elmendorf et al.

TABLE 1. Soil characteristics.

Analyte	Value
gravel (wt %)	7
sand (wt %)	50
silt (wt %)	30
clay (wt %)	13
pH	6.5
water (wt %)	8
water-holding capacity (wt %)	23
ammonia N (ppm)	3.8
nitrate/nitrite N (ppm)	<0.25
total Kjeldahl nitrogen (ppm)	350
total phosphorous (ppm)	338
heterotrophic bacteria (gm^{-1})	2×10^7
oil-degrading bacteria (gm^{-1})	5×10^6

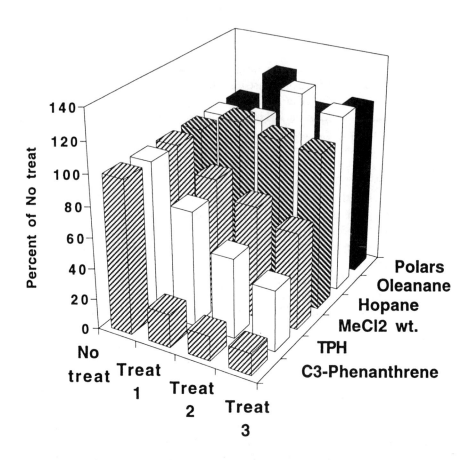

FIGURE 1. The relative degradation of hydrocarbons in aged refinery soil. Concentrations of analytes are normalized to concentrations in samples receiving no nutrient amendments. Polars are the amount of the methylene chloride extractable material that adsorbs to the alumina column used in the cleanup of samples prior to GC/MS analysis, MeCl2 wt. is the weight of methylene chloride extractable material, TPH is total extractable hydrocarbon, estimated by the modified EPA 418.1 method described in the text, and C3P is the sum of the phenanthrene species with three methyl substituents.

1994). There, smaller ring compounds were depleted before larger ones, parents were depleted before alkylated derivatives, and less alkylated species before more alkylated ones. Here, the four-ring pyrenes and chrysenes show greater depletion than the two-ring naphthalenes and three-ring phenanthrenes, and the more highly alkylated naphthalenes and phenanthrenes show greater depletion than the less alkylated ones. Only the alkylated pyrenes and chrysenes show the "conventional" pattern. A possible explanation for these results is that the more readily degradable components were already highly

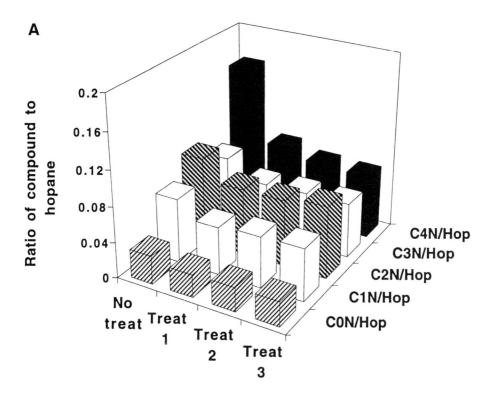

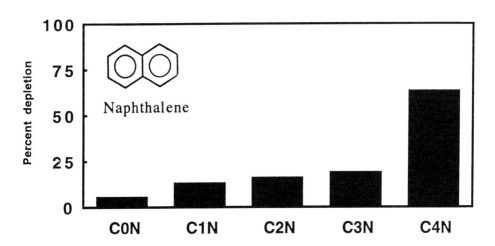

FIGURE 2. (A) The biodegradation, normalized to hopane, of the naphthalenes, and (B) their relative percentage degradation with Treatment 3.

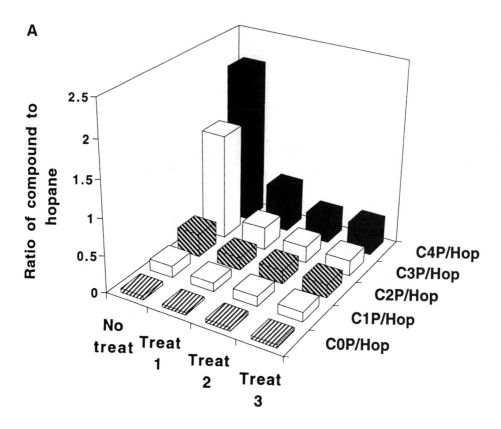

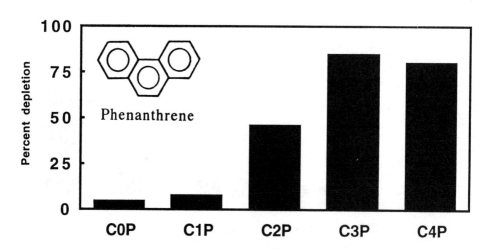

FIGURE 3. (A) The biodegradation, normalized to hopane, of the phenanthrenes, and (B) their relative percentage degradation with Treatment 3.

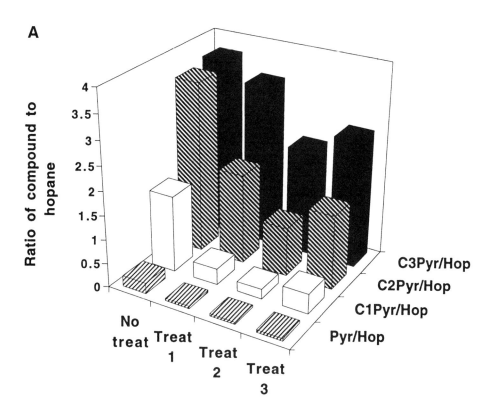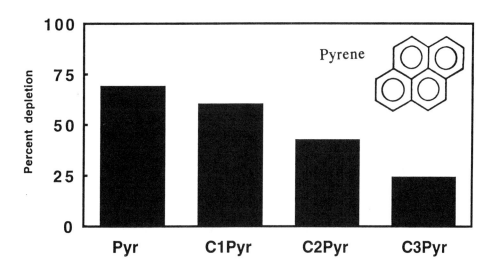

FIGURE 4. (A) The biodegradation, normalized to hopane, of the pyrenes, and (B) their relative percentage degradation with Treatment 3.

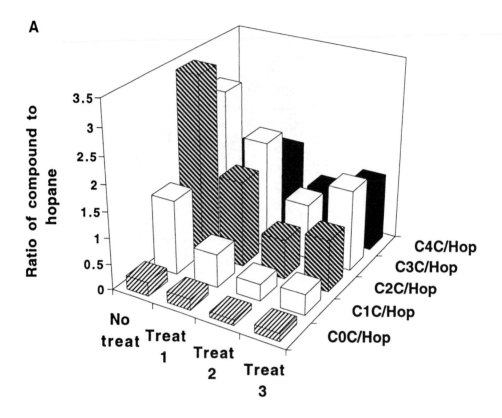

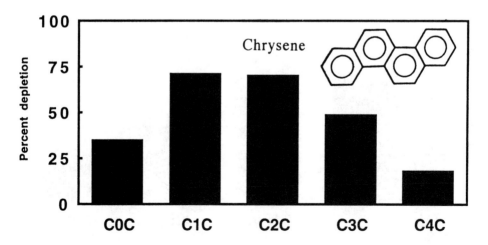

FIGURE 5. (A) The biodegradation, normalized to hopane, of the chrysenes, and (B) their relative percentage degradation with Treatment 3.

biodegraded before the study began, and the remaining portions of these components were associated with the soil in a manner which made them almost unavailable for further biodegradation. Several reports on the effects of aging on PAH biodegradation in soils have appeared recently. PAHs in aged coking plant soil were found to be highly resistant to biodegradation prior to extraction and reapplication to soil (Weissenfels et al. 1992). Phenanthrene was shown to become increasingly resistant to biodegradation upon aging in loam, muck, and aquifer sand (Hatzinger and Alexander 1995). Intrinsic biodegradation rates of two-ring PAHs were found to be significantly lower than those of three- to six-ring PAHs in untreated field soil over a 6-year period (Smith et al. 1994). Various mechanisms suggested for the aging effects include strong adsorption of PAHs onto soil surfaces and/or their diffusion into soil micropores or soil organic matter.

In conclusion, we show that nutrient-stimulated bioremediation is a potentially useful treatment for aged hydrocarbon-contaminated soils; that hopane, oleanane, and polars can be used as conserved internal biomarkers; and that biodegradation patterns of PAHs and their methylated derivatives in aged hydrocarbon-contaminated soils differ from biodegradation patterns of fresh oil in water.

REFERENCES

Atlas, R. M. and R. Bartha. 1992. "Hydrocarbon biodegradation and oil spill bioremediation." *Adv. Microbial Ecol.* 12: 287-338.

Bragg, J. R., R. C. Prince, E. J. Harner, and R. M. Atlas. 1994. "Effectiveness of bioremediation for the Exxon Valdez oil spill." *Nature 368*: 413-418.

Brown, E. J., and J. F. Braddock. 1990. "Sheen screen, a miniaturized most-probable-number method for enumeration of oil-degrading microorganisms." *Appl. Environ. Microbiol. 56*: 3895-3896.

Douglas, G. S., J. K. McCarthy, D. T. Dahlen, J. A. Seavey, W. G. Steinhauer, R. C. Prince, and D. L. Elmendorf. 1992. "The use of hydrocarbon analyses for environmental assessment and remediation." *J. Soil Contam. 1*: 197-216.

Douglas, G. S., R. C. Prince, E. L. Butler, and W. G. Steinhauer. 1994. "The use of internal chemical indicators in petroleum and refined products to evaluate the extent of biodegradation." In R. E. Hinchee, B. C. Alleman, R. E. Hoeppel, and R. N. Miller (Eds.), *Hydrocarbon Remediation.* Lewis Publishers, Boca Raton, FL. pp. 219-236.

Elmendorf, D. E., C. E. Haith, G. S. Douglas, and R. C. Prince. 1994. "Relative rates of biodegradation of substituted polycyclic aromatic hydrocarbons." In R. E. Hinchee, A. Leeson, L. Semprini, and S. K. Ong (Eds.), *Bioremediation of Chlorinated and Polycyclic Aromatic Hydrocarbon Compounds.* Lewis Publishers, Boca Raton, FL. pp. 188-202.

Hatzinger, P. B., and M. Alexander. 1995. "Effect of Aging of Chemicals in Soil on Their Biodegradability and Extractability." *Envtl. Sci. and Tech. 29*(2): 537-545.

Klute, A. (Ed.) 1986. *Methods of Soil Analysis.* Parts 1 and 2, 2nd ed. Soil Science Society of America, Madison, WI.

Landrum, P. F., B. J. Eadie, and W. R. Faust. 1992. "Variation in the bioavailability of polycyclic aromatic hydrocarbons to the amphipod *Diporeia* (spp.) with sediment aging." *Environ. Toxicol. Chem. 11*: 1197-1208.

Leahy, J. G., and R. R. Colwell. 1990. "Microbial degradation of hydrocarbons in the environment." *Microbiol. Revs. 54*: 305-315.

Peters, K., and J. M. Moldowan. 1993. *The Biomarker Guide; Interpreting molecular fossils in petroleum and ancient sediments.* Prentice-Hall, Englewood Cliffs, NJ.

Prince, R. C. 1993. "Petroleum spill bioremediation in marine environments." *Critical Reviews Microbiology. 19*: 217-242.

Prince, R. C., D. L. Elmendorf, J. R. Lute, C. S. Hsu, C. E. Haith, J. D. Senius, G. J. Dechert, G. S. Douglas, and E. L. Butler. 1994. "17a(H),21b(H)-hopane as a conserved internal marker for estimating the biodegradation of crude oil." *Environ. Sci. Technol. 28*: 142-145.

Smith, J. R., R. M. Tomicek, P. V. Swallow, R. L. Weightman, D. V. Nakles, and M. Helbling. 1994. "Definition of Biodegradation Endpoints for PAH Contaminated Soils Using a Risk-Based Approach." In *Abstracts:Proceedings of the Ninth Annual Conference on Contaminated Soils.* Amherst, MA.

U.S. EPA (1983) *Methods for chemical analysis of water and wastes.* U.S. Environmental Protection Agency, Washington DC.

Weissenfels, W. D., H. J. Klewer, and J. Langhoff. 1992. "Adsorption of polycyclic aromatic hydrocarbons by soil particles: influence on biodegradability and biotoxicity." *Appl. Microbiol. Biotechnol. 36*: 689-696.

Characterization of Petroleum Biodegradation Patterns in Weathered Contaminated Soils

John T. Novak, David Schuman, and William Burgos

ABSTRACT

Two soils were contaminated by petroleum, then allowed to weather for 2 years. The soils were then subjected to bioremediation using static systems, flowthrough columns, and slurry systems, each with and without nutrient addition. Petroleum in the static system degraded more slowly and less completely, whereas the slurry system was best. In all systems, the fractions represented by gas chromatograph (GC) elution times from 0 to 10 and 10 to 15 min were rapidly degraded. However, for compounds that eluted at greater than 15 min, the slurry system was superior to the column system and little degradation of these materials was evident in the static units.

INTRODUCTION

Bioremediation is an economically attractive method for the cleanup of petroleum-contaminated soils. However, questions remain about the effectiveness and cost of bioremediation of highly weathered petroleum-contaminated soils. Weathered sites, i.e., sites that have been contaminated for a prolonged time and subjected to a variety of physical, chemical, and biological processes, will contain a large percentage of the most recalcitrant compounds. The recalcitrant compounds generally are characterized by their low solubility, low volatility, strong sorption properties, and resistance to biodegradation. Although regulations vary considerably across the United States, remediation goals are frequently defined in terms of residual total petroleum hydrocarbon (TPH) levels in soils. Because of the recalcitrant nature of weathered petroleum products, there are questions about the attainability of many TPH goals by biological treatment.

The objective of the study was to determine the efficacy of bioremediation for the treatment of a weathered petroleum-contaminated soil and to compare the performance of several different types of processes: static systems, columns, and slurry reactors. The degradation patterns of specific fractions of the

TPH matrix were characterized to determine the rate and extent that specific fractions in the petroleum matrix are degraded. The goal was to develop information that could be used to predict the level of TPH reduction that can be achieved by various treatment approaches. The degradation patterns of petroleum in weathered contaminated soils were characterized to determine the rate at which specific TPH fractions degrade. The type of remediation process required to satisfactorily reduce TPH levels also was investigated.

METHODS AND MATERIALS

Three separate reactor studies were conducted to model remediation processes. Sacrificial static microcosms were used to simulate in situ biodegradation. First, 8-mL screw-top test tubes were loaded with 3 g of soil and 3 mL of water, and every other week 0.8 μL of 30% hydrogen peroxide was added to supply oxygen. These units were sacrificed periodically in triplicate, extracted, then analyzed for organics.

Soil columns were used to model a more active remediation process. The soil columns consisted of 30-cm $\times$ 2.5-cm-diameter glass tubes and 80 g of contaminated soil in each column. Water was passed through daily, and periodically soil was collected from the column for petroleum analysis. Finally, slurry reactors represented the most rigorous bioremediation process, where soil was treated in a stirred aerated reactor. Each reactor contained 60 g of soil. These units were stirred with a variable-speed paddle mixer and samples were periodically removed for analysis.

Two different soils were used: Eagle Point soil, a dark-brown sandy loam; and New River soil, a tan clay. The Eagle Point soil was contaminated with a lightweight crude oil, and the New River soil was contaminated with a middle distillate, similar to a kerosene or jet fuel. Both soils had been contaminated approximately 2 years prior to use in this study. They were stored in closed containers at room temperature, and degradation with and without nutrient (N and P) addition was studied.

TPH was analyzed using a methylene chloride extraction, GC analysis technique (Vodgt 1992). The entire TPH spectrum was broken up into 5-min parcels, based on GC elution time. Specific compounds known to constitute some of the larger fractions of petroleum products were analyzed, then used to identify specific peaks and regions on the sample chromatographs.

RESULTS AND DISCUSSION

Both the rate and degree of degradation increased from static microcosms to columns to slurry reactors. It is thought that the greater availability of oxygen and an increase in mass transfer accounted for the differences between systems. Triplicate TPH averages for the Eagle Point soil and each treatment method are presented in Figure 1. All treatment methods showed separate rapid and slow degradation patterns, but these were especially well defined in the columns and slurry reactors.

Comparing the responses for the three systems, two different phenomena are evident. First, the initial rapid phase of degradation increased from static to column to slurry reactors. For the Eagle Point soil with soil nutrients, rapid phase rates were approximately 600 mg/kg/d for the static system, 1,200 mg/kg/d for the column, and 4,000 mg/kg/d for the slurry system. The second difference in the systems is that the TPH concentration remaining after the rapid degradation phase was 32,000 mg/kg, 12,000 mg/kg, and 2,000 mg/kg for static, column, and slurry reactors, respectively. This indicates that highly weathered petroleum can be biologically remediated, but the rate and residual concentration is dependent on the type of reactor or mixing conditions and the availability of nutrients.

It can also be seen from Figure 1 that nutrient addition increased degradation rates slightly, but had a much more dramatic effect on the residual or undegraded TPH. For all three systems, the residual TPH for the Eagle Point soil was reduced an additional 10,000 to 15,000 mg/kg with addition of nutrients.

Static Microcosms

Static microcosms were used to simulate in situ remediation with oxygen and nutrient enhancement. The TPH concentrations in the static microcosms were monitored over 26 weeks to determine degradation rates for each soil, as well as to determine the effects of nutrients on biodegradation. For the Eagle Point soil, a substantial drop in TPH occurred over the first week as seen in Figure 1. Characterization of the degradation as a function of GC elution time, shown in Figure 2a, illustrates that in the first week, TPH reduction is almost entirely due to losses in the 0- to 10-min and 10- to 15-min fractions. After the first week, most degradation was in the 10- to 15-min and 15- to 20-min parcels. The TPH components eluting above 25 min changed little, and not at all after the ninth week. Although the 10- to 15-min and 15- to 20-min fractions were degraded steadily, they remained as the largest percentage of the contaminant petroleum matrix. Similar results were produced for the New River soil (Figure 2b). For both systems, some degradation occurred in the fractions eluting between 25 and 40 min during the first 9 weeks; but, after 9 weeks, little additional degradation was observed.

Columns

Soil columns were used to simulate conditions present when oxygen-enriched water is cycled through contaminated soil. The columns were monitored for TPH over 13 weeks to determine degradation rates and patterns. The Eagle Point columns exhibited the characteristic biphasic degradation curve with rapid initial rates and slower latter rates (Figure 1). All fractions degraded

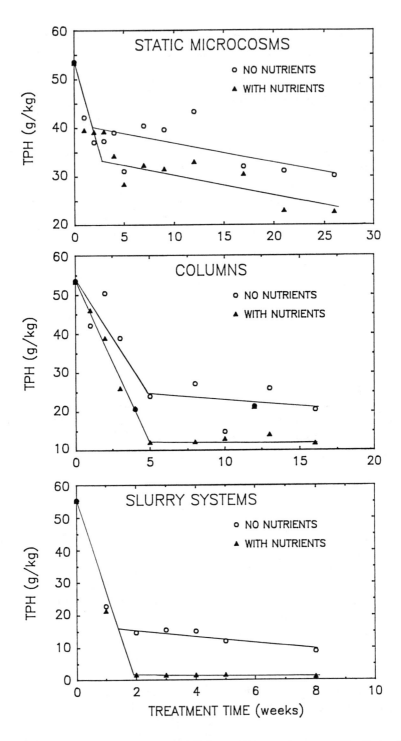

FIGURE 1. The effect of nutrient addition and reactor type on the degradation of petroleum in Eagle Point soil.

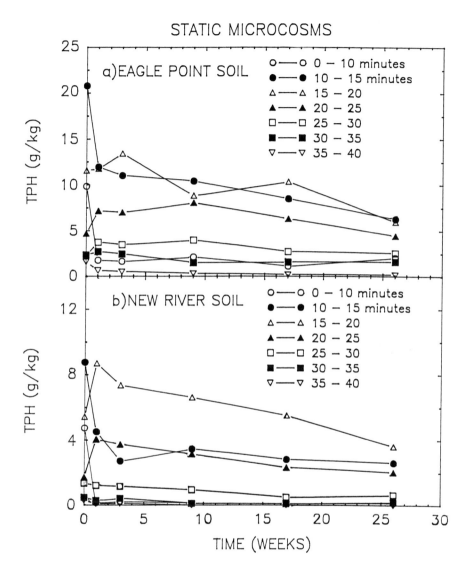

FIGURE 2. **Fractional degradation of petroleum in static microcosms with nutrient addition for (a) Eagle Point soil and (b) New River soil.**

until week 5, as shown in Figure 3. Beyond week 5, all fractions remained relatively unchanged. Figure 3b shows the relative loss of each of the fractions over 13 weeks. The fraction represented by compounds in the 15- to 20-min and 25- to 30-min range remain the least degraded on a percentage basis after 13 weeks. An analysis of the chromatogram indicates that these undegraded fractions comprise specific compounds that resist degradation throughout the treatment cycle. Therefore, after 5 weeks of column treatment, little additional degradation can be expected.

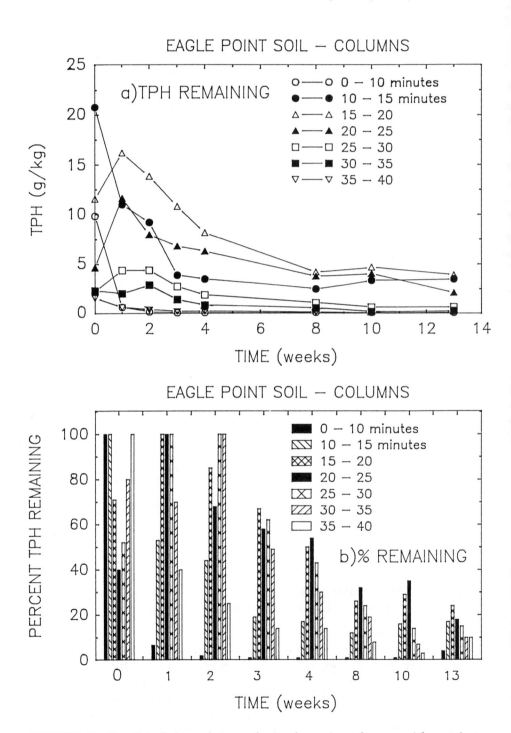

FIGURE 3. Fractional degradation of petroleum in columns with nutrient addition: (a) TPH remaining and (b) percent remaining.

Slurry Reactors

Slurry reactors were used to simulate optimum biodegradation, by providing well-aerated and well-mixed conditions. Degradation patterns for both soils and treatments followed biphasic curves (Figure 1). Nutrients enhanced the degradation rate and reduced the level at which slow degradation begins. Data in Figure 4a for the Eagle Point soil illustrate the relatively complete degradation of all fractions by week 2 for the nutrient-enhanced study. The 15- to 25-min fraction remains as the largest portion of TPH present at the termination of the experiment. The removal rates for the slurry system without nutrients, shown in Figure 4b, are similar over the first week, but slow dramatically thereafter.

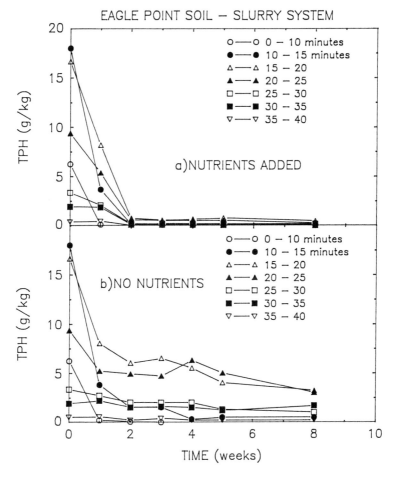

FIGURE 4. Effect of nutrient addition on degradation of petroleum fractions in a slurry system: (a) nutrients added and (b) no nutrients added.

Overall Pattern

The biodegradation patterns of several types of bioremediation systems were studied for two contaminated soils. As expected, degradation effectiveness increased from static microcosms to columns to slurry reactors. Nutrients aided biodegradation for Eagle Point soil under all treatments, whereas for the New River soil, nutrients enhanced only the slurry system.

For the columns and static microcosms, the 0- to 10-min and 10- to 15-min parcels were mainly responsible for the rapid initial TPH reductions. All fractions contributed to the initial rapid degradation in the nutrient-amended slurry reactors. At TPH concentrations present in this study, the nutrient-enhanced slurry reactors appeared to be the only treatment method able to reduce the TPH levels to below 1,000 mg/kg for weathered soil.

Figures 5a and 5b summarize the rapid phase degradation rates for the Eagle Point and New River soils, respectively. It can be seen that degradation

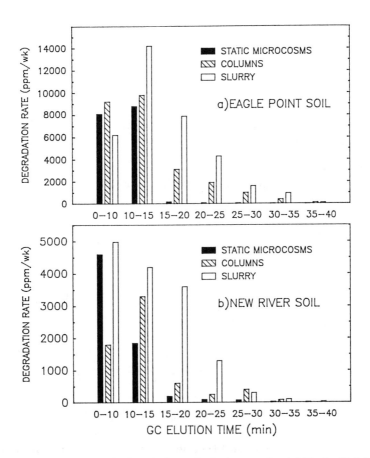

FIGURE 5. Initial (rapid phase) degradation rates for (a) Eagle Point soil and (b) New River soil.

in the static system is comparable to the other systems for compounds in the 0- to 15-min elution range. This range is represented by aliphatics up to 14-carbon and 2-ring aromatics, compounds that are relatively soluble and degradable under aerobics conditions. In the ranges above the 15-min elution time, degradation in the slurry system is at least 20 times faster that in the static system and two to five times faster that in the recirculated flow columns. Also, above the 15-min elution time, each fraction was increasingly more difficult to degrade. However, degradation in all ranges was possible using both the column and slurry systems.

Data presented in Figure 6 show the distribution of ungraded TPH for each reactor type at the end of the rapid degradation phase. These data show that for the column and slurry systems, nearly complete degradation of the 0- to 10-min fraction occurs. The slurry system also degrades the 10- to 15-min fraction much more completely than the column system. Following the rapid phase of degradation, the remaining distributions look remarkably similar for all systems, with the majority of undegraded compounds in the 15- to 20-min and 20- to 25-min range. These compounds appear to be relatively undegradable by any treatment method.

It is clear that the success of any biological remediation scheme depends on the type of contamination and the contaminant level. For the light fractions,

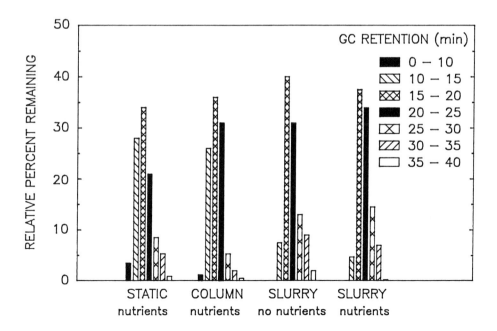

FIGURE 6. Distribution of remaining fraction in all reactor types following the rapid degradation phase.

oxygen and nutrient addition appear to be adequate to promote degradation, and in situ treatment is feasible. For the least soluble materials, both the column and slurry systems will reduce TPH. Nutrients appear to be important in determining the degree to which the organics will be degraded. The types of compounds degraded as shown in Figure 6 for the slurry system with and without nutrients are unaffected by the presence or absence of nutrients; only the extent of degradation differs.

REFERENCE

Vodgt, J. 1992. "Bioremediation of Petroleum Hydrocarbon Contaminated Soil." M.S. Thesis. Virginia Polytechnic Institute and State University, Blacksburg, VA.

Overview of Biomolecular Methods for Monitoring Bioremediation Performance

Fred J. Brockman

ABSTRACT

Determining the success of in situ bioremediation is complicated by a number of factors, including assembling a contaminant mass balance at field sites, distinguishing biotic from abiotic loss of contaminant, and proving that biodegradative-potential assays and culture-based enrichments and enumerations performed in the laboratory provide accurate information on what is occurring in situ. Biomolecular methods are now being used to characterize the nucleic acids or cell membranes of the microbial community in the environmental sample. The major advantage of these molecular characterization methods is that the analyses are direct, whereas analyses performed on any kind of laboratory incubation are indirect and involve artifactual changes in the microbial community structure and metabolic activity. Because samples are immediately frozen after acquisition, direct molecular methods preserve the in situ metabolic status and microbial community composition. In addition, direct extraction of nucleic acids or cell membranes from environmental samples accounts for the very large proportion of microorganisms (90 to 99.9%) that are not readily cultured in the laboratory but may be responsible for the majority of the biodegradative activity of interest. These biomolecular methods complement the classic microbiological methods and provide for a more comprehensive and defensible interpretation of the in situ microbial community and its response to both engineered bioremediation systems and natural attenuation processes.

SCOPE

The objectives of this overview are to briefly describe the information obtained from nucleic acid-based and cell-membrane-based molecular methods, and to summarize research that has used these methods for analysis of environmental samples and samples taken during site bioremediation.

NUCLEIC ACID (DNA AND RNA)-
BASED METHODS

How Do They Work?

All nucleic acid-based methods are predicated upon chemical extraction and purification of cellular nucleic acids from the microorganisms present in the soil, sediment, or water sample. DNA contains the information for the production of all cellular constituents, with a specific gene(s) in the DNA encoding for each cellular constituent. When the cell requires the production of a specific cellular constituent (such as a biodegradative enzyme), RNA is synthesized from the gene and used to direct the production of the needed cellular constituent. For example, trichloroethylene (TCE) is degraded by the soluble methane monooxygenase (sMMO) in many methanotrophic microorganisms. Nucleic acids extracted from the microbial community in the environmental sample are analyzed for the presence and relative abundance of the sMMO DNA (gene) or sMMO messenger RNA (mRNA or transcript). The relative abundance of the sMMO gene provides information on the methanotrophic TCE biodegradative potential, but does not tell us whether the microorganisms are producing the sMMO enzyme. Strong evidence that the sMMO enzyme is being produced in situ is obtained by demonstrating the presence of the sMMO transcript in the sample.

All nucleic acid-based methods involve the addition of single-stranded nucleic acid sequences specific for a particular gene to (single-stranded) DNA or RNA extracted from the environmental sample. The added nucleic acid is then allowed to pair or anneal with identical and very similar environmental nucleic acid sequences to form double-stranded hybrids. For hybridization approaches, the added nucleic acid (termed a probe) corresponds to a gene or transcript and is labeled with a chemical. The amount of chemical that is incorporated into the hybrid reports on the density of the gene or transcript in the environmental sample. The lower detection limit is 10^5 to 10^6 gene copies per g soil or sediment. A second major type of approach involves the addition to the environmental nucleic acid of two much shorter DNA sequences (termed primers, which are specific for a biodegradative gene or transcript) and materials enabling DNA synthesis.

The sample is then exposed to conditions (termed the polymerase chain reaction, PCR) that allow any resulting hybrids to be amplified by many orders of magnitude, allowing the detection and analysis of genes or transcripts that are present at low concentrations. The lower detection limit for PCR in soil or sediment extracts is usually 10^2 to 10^3 copies per g. Variations of the PCR include arbitrarily primed PCR (AP-PCR; also termed randomly amplified polymorphic DNA [RAPD] analysis) and repetitive DNA-PCR (rep-PCR). These techniques use a single primer, which does not target a specific gene or transcript, and allow the generation of highly specific DNA fingerprints that can be used to rapidly monitor specific microbial populations in the environment.

Validation of Methods on Environmental Samples

Numerous microcosm and field studies have demonstrated the utility of nucleic acid-based methods for characterizing microbial populations and their response to chemical contaminants. The following two sections focus on research that targeted analysis of DNA and RNA, respectively.

Genes as the Target. Early studies involved hybridizing DNA probes to individual colonies of environmental bacteria or most-probable-number (MPN) enrichments to estimate the number of microorganisms possessing the gene and comparing the results to measures of biodegradative activity. Correlations were found between the percentage of the total colony-forming units (CFUs) that hybridized to the TOL (toluene) and NAH7 (naphthalene) catabolic plasmids, and mineralization of polycyclic aromatic hydrocarbons (PAHs) in microcosms spiked with synthetic oil (Sayler et al. 1985); between MPN values for 3-chlorobenzoate (3-CB)-degraders using a 3-CB gene probe, and the V_{max} for 3-CB uptake in flowthrough freshwater and sediment microcosms spiked with 3-CB (Fulthorpe and Wyndham 1989); and between population estimates using a 2,4- dichlorophenoxyacetic acid (2,4-D) gene probe, an MPN from 2,4-D enrichments, and 2,4-D degradation rates in spiked water samples (Amy et al. 1990).

Later studies used direct extraction of DNA from environmental samples and compared gene probe analysis with a measurement of biodegradative activity and/or an estimate of the population of culturable contaminant-degrading microorganisms. Correlations were obtained between the density of mercury (Hg)-resistance genes and the induction of biologic Hg reduction in Hg contaminated and uncontaminated pond water (Barkay et al. 1991), between hybridization with a naphthalene gene probe and culturable naphthalene degraders in soil microcosms from a PAH-contaminated site amended with salicylate (an inducer of naphthalene degradation) (Ogunseitan and Olson 1993), and between estimated populations of 2,4-D degraders using DNA probes (for two pathways of 2,4-D degradation) and MPN values from 2,4-D enrichments (Holben et al. 1992, Ka et al. 1994).

Recently, PCR has been used to detect the naphthalene dioxygenase gene in coal tar-contaminated surface soil and in uncontaminated near-surface sediment (Herrick et al. 1993), to demonstrate the presence of the 2,3-dihydroxybiphenyl dioxygenase gene in polychlorinated biphenyls (PCB)-contaminated river sediments (Erb and Wagner-Dobler 1993), to detect catabolic genes associated with a 3-CB-degrading bacterium 14 months after it had been injected into a contaminated aquifer (Thiem et al. 1994), and to characterize the distribution of a naphthalene-degrading bacterium after transport through a soil block (Palumbo et al. 1995). The transport of four different bacterial strains introduced into an aquifer has been characterized using PCR primers directed at randomly cloned sequences that were not present in the site microflora

(Burlage et al. 1995). In addition, AP-PCR has been used in soil field plots to demonstrate the ability to track an introduced, highly efficient PCB-degrading bacterium (R.J. Steffan, unpublished data).

Transcripts as the Target. mRNA from the naphthalene dioxygenase gene was detected 1 month after microcosms containing soil from a PAH-contaminated site were seeded with naphthalene-degrading cells isolated from the site and amended with the inducer salicylate (Ogunseitan et al. 1991). In similar microcosms that contained only indigenous microbiota, the salicylate concentration that yielded the greatest amount of naphthalene dioxygenase mRNA also showed the greatest mineralization of naphthalene (Ogunseitan and Olson 1993). A hybridization approach termed the ribonuclease protection assay (RPA) was used to demonstrate a correlation between naphthalene dioxygenase mRNA concentrations, soluble naphthalene concentrations, and naphthalene mineralization rates in microcosms containing soils from various PAH-contaminated sites (Fleming et al. 1993). Mercuric reductase mRNA has been quantitated under field environmental conditions in lake and river water (Jeffrey et al. 1994), and a companion study showed that the abundance of mercuric reductase transcripts was correlated to community heterotrophic activity (Nazaret et al. 1994).

Use of Nucleic Acid-Based Methods at Field Sites Undergoing Bioremediation

Oil- and Diesel-Contaminated Sites. Feasibility studies for in situ bioremediation were performed at seven gasoline-contaminated sites (Samson et al. 1994). Populations of catechol 2,3-dioxygenase-positive and alkane hydroxylase-positive colonies determined by gene probes were correlated to the mineralization of the test substrates toluene, benzene, and hexadecane in groundwater samples at several of the sites. However, concentrations of toluene, ethylbenzene, and xylenes (BTEX) and total petroleum hydrocarbons (TPH) were correlated to the extent of mineralization and the percentage of probe-positive colonies at only one of the seven sites.

The same probes were used to monitor the treatment of oil- and diesel-contaminated soil in aboveground heap piles (McNicoll et al. 1994). Each pile contained 900 tons (816,480 kg) soil with a network of aeration piping and a nutrient delivery system. The piles were monitored from May to November, yielding a 97% decrease in TPH, with 99% of the reduction due to biological degradation. The proportion of the total colonies that were probe-positive was correlated to increasing soil temperature and to the hexadecane mineralization rate (C. Greer, unpublished data).

Gasoline-contaminated Sites. Samples from four aquifers undergoing bioventing were studied using an MPN format in microdilution plates supplemented with the (TOL plasmid-encoded) toluene pathway intermediate *meta*-toluate. A whole TOL plasmid probe was hybridized to DNA extracted from the enrichments to estimate densities of TOL plasmid-containing bacteria able to degrade toluene and xylenes. Bioventing commonly increased the numbers of TOL plasmid-containing bacteria by two to three orders of magnitude, as compared to nearby portions of the contaminated sites that were not biovented (J. D. Randall and B.B. Hemmingsen, unpublished data).

Jet Propellant (JP-5)-Contaminated Site. The distribution of the naphthalene dioxygenase/polyaromatic hydrocarbon dioxygenase, catechol 2,3-dioxygenase, and alkane hydroxylase genes was characterized by PCR (Chandler and Brockman 1995). Samples were taken before and after short-term and long-term bioventing. The naphthalene dioxygenase/polyaromatic hydrocarbon dioxygenase and catechol 2,3-dioxygenase genes were present in nearly all samples, and were positively correlated with TPH and BTEX concentrations. Although ^{14}C-dodecane was mineralized in samples from all locations, the alkane hydroxylase gene was rarely detected, indicating that other alkane biodegradative genes were responsible for JP-5 degradation at the site. At a given location, aerobic acetate mineralization increased after bioventing, but the acridine orange direct count (AODC) and densities of the target genes were approximately the same before and after bioventing. This result suggested that enhanced transcription or enzymatic activity may be a more important response to bioventing than changes in community structure, or that intrinsic bioremediation processes were well established before the engineered remediation was initiated.

TCE-Contaminated Site. Stimulation of methanotrophic populations for the purpose of TCE bioremediation was conducted at the Savannah River Site, South Carolina, using horizontal wells at depths of 175 ft (53 m; injection well in saturated zone) and 80 ft (24 m; extraction well in vadose zone) below the surface. Gene probes were used to analyze nucleic acids extracted from sediment samples before the injection of air; after 21 weeks of air injection; after 15 weeks of methane (1% in air) injection; after 25 additional weeks of methane (4% in air) injection; and after 12 weeks of methane (4% in air), nitrous oxide (N), and triethyl phosphate (P) injection. The sMMO and methanol dehydrogenase gene probes were used to target methanotrophic populations. The toluene dioxygenase, toluene-4-monooxygenase, and ammonia monooxygenase gene probes correspond to enzymes that cometabolically degrade TCE.

Bioremediation performance was also assessed by measuring TCE concentrations in soil gas, groundwater, sediment, and the extraction well; methanotrophic MPN and biodegradative potential in groundwater and sediment samples; TCE degradation kinetics in column experiments and a differential soil bioreactor; and numerical simulations that used the data to model the bioremediation process. A comprehensive summary of these studies and the performance of the demonstration has been published (Hazen et al. 1994). The simulations showed that the TCE degradation rate at the site during the campaigns was in the order pulse 44% CH_4/continuous NP > 1% CH_4 > 4% CH_4 > pulse CH_4 (Travis and Rosenberg 1994).

The gene probes showed four patterns of response during the demonstration; data are from the laboratories of A. Ogram and G. Saylet (Brockman et al. 1994). Sequences homologous to the sMMO and methanol dehydrogenase gene probes, and methanotrophic MPN values were the highest at the end of the 1% CH_4 campaign and pulse 4% CH_4/continuous NP campaign. Thus, the frequency of detection with methanotroph-targeted gene probes, the methanotrophic MPN, and the TCE biodegradation rate at the site were all correlated. Sequences homologous to the toluene dioxygenase (*C1* gene) gene probe were detected at high frequencies prior to methane injection, but the frequency declined dramatically as methane injection progressed. In contrast, the frequency of detection of sequences homologous to the toluene-4-monooxygenase gene probe increased throughout the demonstration. The ammonia monooxygenase gene probe did not detect homologous sequences in any of the samples.

The RPA provided direct evidence, in several samples from the pulse 4% CH_4/continuous NP campaign, for sMMO and/or toluene dioxygenase gene expression under in situ field conditions (Gregory et al. 1995). This is the first documented detection of mRNA from catabolic genes in deep sediments under in situ field conditions.

Comparison of the cultural MPN method with the gene probe and RPA results suggested that the cultural method underestimated the methanotrophic biomass in many of the sediment samples by several orders of magnitude (Brockman et al. 1994). The underestimate by the cultural method may have been caused by the inability to disrupt microcolonies or to sufficiently culture methanotrophs so that growth could be visibly detected. Therefore, nucleic acid-based methods were important tools for assessing bioremediation performance at the site.

CELL MEMBRANE-BASED METHODS

How Do They Work?

Cell membrane-based methods involve quantitative extraction of microbial lipids from microbial cell membranes by adding solvent to the environmental sample. The purified lipids are fractionated into various classes and then derivitized to allow analysis of their chemical structure by gas chromatography-mass spectrometry (GC-MS). The chemical structures of microbial lipids have been found to provide a wealth of information on the community structure, nutritional status, and metabolic activity of soil and subsurface microbial communities (reviewed in Tunlid and White, 1992). For example, signature lipid biomarkers (SLBs) in the phospholipid fatty acids (PLFAs) can identify dominant physiological types of microorganisms in communities and identify metabolic stress, the nutritional status of the community can be determined by examining the polyhydroxyalkanoate (PHA) to PLFA ratio, and dominant terminal electron-accepting processes can be determined by characterizing the quinone biomolecules. These methods are complementary to the nucleic acid-based methods because they provide certain types of process-level physiological information more readily than does nucleic acid analysis. The lower detection limit for characterization of microbial lipids is approximately 10^7 cells.

Validation of Methods on Environmental Samples

In soil columns amended with natural gas and TCE, the TCE degradation was correlated with the presence of specific methane-oxidizing bacteria and the 18: lw8c SLB (Nichols et al. 1987). In subsequent studies investigating TCE degradation, sediments and soils exposed to alkanes in column and reactor experiments were compared to the identical material that was not exposed to the alkanes. Exposure to propane resulted in high levels of the unusual SLBs 16: lw6c and 10Me18:0, and propane-oxidizing bacterial isolates from the samples also contained high levels of these SLBs (Ringelberg et al. 1989). Similarly, exposure to methane resulted in high levels of the SLB 18: l w8c, with methane-oxidizing bacterial isolates containing high levels of this SLB. The correspondences validated the utility of cell membrane biomarkers for defining certain dominant physiological types of microorganisms in subsurface samples. More recently, specific anaerobic bacteria capable of dehalogenating chlorinated compounds have been detected in soils by their unusual lipopolysaccharide fatty acids (Ringelberg et al. 1994).

Use of Cell Membrane-Based Methods at Field Sites Undergoing Bioremediation

SLB analyses were performed on samples from the TCE bioremediation demonstration at the Savannah River Site, South Carolina, to monitor the increase in methanotrophic biomass upon injection of methane into the subsurface. In groundwater samples, SLB indicative of type II methanotrophs increased from 4 to 10% of the total PLFA before methane addition to 20 to 90% of the total PLFA after methane addition (Pfiffner et al. 1994).

CONCLUSION

Over the past decade, microcosm and field studies have demonstrated the utility of biomolecular methods for characterizing microbial communities. Biomolecular methods are now being successfully applied for monitoring the performance of engineered field bioremediation systems. These direct methods preserve the in situ metabolic status and microbial community composition, and account for the very large proportion of microorganisms that are not readily cultured in the laboratory. The detection and quantitation of specific biodegradative genes, the corresponding mRNA indicative of enzyme production, signature lipid biomarkers, and other information containing lipids are now valuable molecular diagnostic tools for the initial characterization of sites and for monitoring bioremediation performance.

REFERENCES

Amy, P. S., M. V. Staudaher, and R. J. Seidler. 1990. "Comparison of a Gene Probe With Classical Methods for Detecting 2,4-Dichlorophenoxyacetic Acid (2,4-D)-Biodegrading Bacteria in Natural Waters." *Current Microbiology* 21: 95-101.

Barkay, T., R. R. Turner, A. VandenBrook, and C. Liebert. 1991. "The Relationship of Hg(II) Volatilization From a Freshwater Pond to the Abundance of *mer* Genes in the Gene Pool of the Indigenous Microbial Community." *Microbial Ecology* 21: 151-161.

Brockman, F. J., J. P. Bowman, J. T. Fleming, I. Gregory, A. Ogram, G. S. Sayler, W. Sun, and D. Zhang. 1994. "Nucleic Acid Technology Report for the In Situ Bioremediation Demonstration (Methane Biostimulation) of the Savannah River Integrated Demonstration Project DOE/OTD." Draft Technical Report, Pacific Northwest Laboratory, Richland, WA.

Burlage, R. S., A. V. Palumbo, and J. McCarthy. 1995. "DNA Extraction of Bacterial Strains in a Field Transport Experiment: Detection Using PCR Amplification." Submitted, *Molecular Ecology*.

Chandler, D. P., and F. J. Brockman. 1995. "Qualitative MPN-PCR to Estimate Biodegradative Gene Numbers at a Jet Fuel (JP-5) Contaminated Site." Submitted, *Applied Biochemistry and Biotechnology*.

Erb, R. W., and I. Wagner-Dobler. 1993. "Detection of Polychlorinated Biphenyl Degradation Genes in Polluted Sediments by Direct DNA Extraction and Polymerase Chain Reaction." *Applied and Environmental Microbiology* 59: 4065-4073.

Fleming, J. T., J. Sanseverino, and G. S. Sayler. 1993. "Quantitative Relationship Between Naphthalene Catabolic Gene Frequency and Expression in Predicting PAH Degradation in Soils at Town Gas Manufacturing Plants." *Environmental Science and Technology* 27: 1068-1074.

Fulthorpe, R. R., and R. C. Wyndham. 1989. "Survival and Activity of a 3-Chlorobenzoate-Catabolic Genotype in a Natural System." *Applied and Environmental Microbiology* 55: 1584-1590.

Gregory, I., J. P. Bowman, L. Jimenez, J. T. Fleming, S. M. Pfiffner, and G. S. Sayler. 1995. "Cometabolic Gene Distribution and Expression in a Trichloroethylene-Contaminated Site Undergoing In Situ Bioremediation." Submitted, *Applied and Environmental Microbiology*.

Hazen, T. C., K. H. Lombard, B. B. Looney, M. Z. Enzien, J. M. Dougherty, C. B. Fliermans, J. Wear, and C. A. Eddy-Dilek. 1994. "Summary of In-Situ Bioremediation Demonstration (Methane Biostimulation) Via Horizontal Wells at the Savannah River Site Integrated Demonstration Project." In G. W. Gee and N. R. Wing (Eds.), *In Situ Remediation: Scientific Basis for Current and Future Technologies*, pp. 137-150. Battelle Press, Richland, WA.

Herrick, J. B., E. L. Madsen, C. A. Batt, and W. C. Ghiorse. 1993. "Polymerase Chain Reaction Amplification of Naphthalene-Catabolic and 16S rRNA Gene Sequences From Indigenous Sediment Bacteria." *Applied and Environmental Microbiology* 59: 687-694.

Holben, W .E., B. M. Schroeter, V. G. M. Calabrese, R. H. Olsen, J. K. Kukor, V. O. Biederbeck, A. E. Smith, and J. M. Tiedje. 1992. "Gene Probe Analysis of Soil Microbial Populations Selected by Amendment With 2,4-Dichlorophenoxyacetic Acid." *Applied and Environmental Microbiology* 58: 3941-3948.

Jeffrey, W. H., S. Nazaret, and R. von Haven. 1994. "Improved Method for Recovery of mRNA From Aquatic Samples and its Application to Detection of *mer* Expression." *Applied and Environmental Microbiology* 60:1814-1821.

Ka, J. O., W. E. Holben, and J. M. Tiedje. 1994. "Use of Gene Probes to Aid in Recovery and Identification of Functionally Dominant 2,4-Dichlorophenoxyacetic Acid-Degrading Populations in Soil." *Applied and Environmental Microbiology* 60:1116-1120.

McNicoll, D. M., A. S. Baweja, M. J. L. Robin, C. W. Greer, and F. D'Addario. 1994. "Operation, Monitoring and Performance of a Bioreactor Engineered to Treat Soils Containing Petroleum Hydrocarbons." In *Proceedings of GASReP Symposium 4*, pp. 545-556.

Nazaret, S., W. H. Jeffrey, E. Saouter, R. von Haven, and T. Barkay. 1994. "*merA* Gene Expression in Aquatic Environments Measured by mRNA Production and Hg(II) Volatilization." *Applied and Environmental Microbiology* 60: 4059-4065.

Nichols, P. D., J. M. Henson, C. P. Antworth, J. Parsons, J. Wilson. and D. C. White. 1987. "Detection of a Microbial Consortia Including Type II Methanotrophs by Use of Phospholipid Fatty Acids in Aerobic Halogenated Hydrocarbon Degrading Soil Column Enriched With Natural Gas." *Environmental Toxicology and Chemistry* 6: 89-97.

Ogunseitan, O. A., I. L. Delgado, Y.-L. Tsai, and B. H. Olson. 1991. "Effect of 2-Hydroxybenzoate on the Maintenance of Naphthalene-Degrading Pseudomonads in Seeded and Unseeded Soil." *Applied and Environmental Microbiology* 57: 2873-2879.

Ogunseitan, O. A., and B. H. Olson. 1993. "Effect of 2-Hydroxybenzoate on the Rate of Naphthalene Mineralization in Soil." *Applied Microbiology and Biotechnology* 38: 799-807.

Palumbo, A. V., J. McCarthy, P. Jardine, R. S. Burlage, and G. W. Wilson. 1995. "Transport of Bacteria in an Unsaturated Soil Block." Submitted, *Environmental Science and Technology.*

Pfiffner, S. M., D. B. Ringelberg, D. C. White, T. J. Phelps, A.V. Palumbo, J. P. Bowman, G. S. Sayler, F. J. Brockman, J. Sinclair, M. Enzien, J. Dougherty, M. Franck, J. Wear, C. Fliermans, and T. C. Hazen. 1994. "Community Structure Report for the In Situ Bioremediation Demonstration (Methane Biostimulation) of the Savannah River Integrated Demonstration Project DOE/OTD." Draft Technical Report, Oak Ridge National Laboratory, Oak Ridge, TN.

Ringelberg, D. B., J. D. Davis, G. A. Smith, S. M. Pfiffner, P. D. Nichols, J. S. Nickels, J. M. Henson, J. T. Wilson, M. Yates, D. H. Kampbell, H. W. Read, T. T. Stocksdale, and D. C. White. 1989. "Validation of Signature Polarlipid Fatty Acid Biomarkers for Alkane-Utilizing Bacteria in Soils and Subsurface Aquifer Materials." *FEMS Microbiology Ecology* 62: 39-50.

Ringelberg, D. B., T. Townsend, K. A. DeWeerd, J. M. Suflita, and D. C. White. 1994. "Detection of the Anaerobic Dechlorinator *Desulfomonile tiedjei* in Soil by its Signature Lipopolysaccharide Branched-Long-Chain Hydroxy Fatty Acids." *FEMS Microbiology Ecology* 14: 9-18.

Samson, R., C. W. Greer, J. Hawari, N. Matte, D. Beamier, C. Beaulieu, D. Millette, and A. Pilon. 1994. "Biotreatability Assessment for In-Situ Treatment of Sites Impacted With Light Hydrocarbons." In *Proceedings of GASReP Symposium 4*, pp. 225-252.

Sayler, G. S., M. S. Shields, E. T. Tedford, A. Breen, S. W. Hooper, K. M. Sirotkin, and J. W. Davis. 1985. "Application of DNA-DNA Hybridization to the Detection of Catabolic Genotypes in Environmental Samples." *Applied and Environmental Microbiology* 49: 1295-1303.

Thiem, S. M., M. L. Krumme, R. L. Smith, and J. M. Tiedje. 1994. "Use of Molecular Techniques to Evaluate the Survival of a Microorganism Injected Into an Aquifer." *Applied and Environmental Microbiology* 60: 1059-1067.

Travis, B. J., and N. D. Rosenberg. 1994. *Numerical Simulations in Support of the In Situ Bioremediation Demonstration at Savannah River.* Technical Report No. LA-12789-MS, Los Alamos National Laboratory, Los Alamos, NM.

Tunlid, A., and D. C. White. 1992. "Biochemical Analysis of Biomass, Community Structure, Nutritional Status, and Metabolic Activity of the Microbial Communities in Soil." In G. Stotzky and J.-M. Bollag (Eds.), *Soil Biochemistry*, pp. 229-262. Marcel Dekker, Inc., New York, NY.

Phospholipid Analysis of Extant Microbiota for Monitoring In Situ Bioremediation Effectiveness

Holly C. Pinkart, David B. Ringelberg,
Julia O. Stair, Susan D. Sutton,
Susan M. Pfiffner, and David C. White

ABSTRACT

Two sites undergoing bioremediation were studied using the signature lipid biomarker (SLB) technique. This technique isolates microbial lipid moieties specifically related to viable biomass and to both prokaryotic and eukaryotic biosynthetic pathways. The first site was a South Pacific atoll heavily contaminated with petroleum hydrocarbons. The second site was a mine waste reclamation area. The SLB technique was applied to quantitate directly the viable biomass, community structure, and nutritional/physiological status of the microbiota in the soils and subsurface sediments of these sites. All depths sampled at the Kwajalein Atoll site showed an increase in biomass that correlated with the co-addition of air, water, and nutrients. Monoenoic fatty acids increased in abundance with the nutrient amendment, which suggested an increase in gram-negative bacterial population. Ratios of specific phospholipid fatty acids indicative of nutritional stress decreased with the nutrient amendment. Samples taken from the mine reclamation site showed increases in total microbial biomass and in *Thiobacillus* biomass in the plots treated with lime and bactericide, especially when a cover soil was added. The plot treated with bactericide and buffered lime without the cover soil showed some decrease in *Thiobacillus* numbers, but was still slightly higher than that observed in the control plots.

INTRODUCTION

Bioremediation of contaminated sites offers the most cost-effective means of protecting groundwater resources. A major limitation to the application of bioremediation has been an inability to predict and monitor the effects of specific treatments on microbial communities when using classical microbiology techniques such as plate count or most probable number (MPN) enumerations. An alternative to the classical techniques is the signature lipid biomarker (SLB) technique. This paper addresses two sites where the SLB technique was used to monitor both bioremediation and bioreclamation efforts.

Kwajalein Atoll is an island in the South Pacific that has been contaminated with petroleum hydrocarbons following years of use by the military. This site was chosen for bioremediation following feasibility studies showing that the soil conditions, temperature, and indigenous microbiota favored this type of approach. For the in situ treatments, injection wells were used to deliver combinations of air, water, and nutrients. Two plots were used as controls and received no amendments, and ten plots received amendments. Sediments were sampled from both control plots and amended plots at depths of 4 and 5 ft (1.2 and 1.5 m). Six ex situ plots also were sampled. These plots were polypropylene-lined cells that contained excavated soils from the contaminated areas. The ex situ plots were treated with the same amendments as the in situ plots.

The second site is a semi-arid mine waste area undergoing bioreclamation activities. This site contained acidified soil resulting from microbial activities attributed to *Thiobacillus*. The reclamation approach was to inhibit further microbial activities through the use of lime and a bactericide. The amount of lime applied to treated plots was determined using the SMP buffer test (Shoemaker et al. 1961), and the amount of ProMac (bactericidal surfactant) was applied according to the manufacturer's recommendations. The layout of the control and treated plots is shown in Figure 1.

Lipid analysis provides a quantitative means to measure total viable microbial biomass, community structure, and nutritional status of microbial communities over the course of bioremediation activities. Phospholipid fatty acid (PLFA) analysis is based on the following: (1) phospholipids are indicative of viable biomass; (2) phospholipid fatty acids are synthesized via pathways specific to different microbial groups allowing for the establishment of SLBs; and (3) ratios of specific phospholipid fatty acids can be used to assess physiological status (Vestal and White 1989, Guckert et al. 1986). Because lipids can be directly extracted from groundwater and soil matrices (without culturing or isolating the microbes), this method has proved very useful for monitoring the effectiveness of bioremediation efforts at sites contaminated with trichloroethylene (TCE) and tetrachloroethylene (PCE) (Phelps et al. 1988, 1990, 1991; Nichols et al. 1987), polychlorinated biphenyls (PCBs) (Hill et al. 1989), and linear alkanosulfonates (LAS) (Federle et al. 1991). The SLB analysis has recently been strengthened with the discovery that the lipid extraction yields DNA suitable for gene probing (Kehrmeyer et al. 1995), which allows for the detection of specific genes involved in biodegradation.

METHODS

The Kwajalein site was treated in situ with the addition of air water and nutrients. The ex situ samples had been excavated from the site, tranferred to a plastic-lined trough, and subjected to the same treatments as the in situ plots.

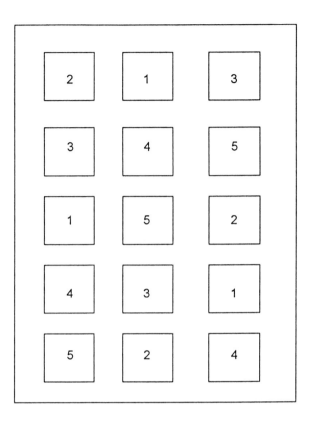

	Treatment 1	Treatment 2	Treatment 3	Treatment 4	Treatment 5	
		Coversoil	Coversoil	Lime* and Bactericide	Coversoil	0'
						1'
	Control	Lime* and Bactericide	Lime*		Excess** Lime	
						2'
						3'
						4'

* Represents the amount of lime required to neutralize the mine waste to pH 7.0
** Represents the total lime required to neutralize active and potential acidity

FIGURE 1. Experimental design for the mine waste reclamation site. Upper figure is a schematic representation of the layout of amended and control plots. Plot number corresponds to the treatment scheme shown in the lower plot.

The mine waste reclamation site was set up as shown in Figure 1. The number on each plot refers to the treatment applied to that area. The experimental design is shown at the bottom of the figure. Samples were collected from the cover soil region, from the amended region, and from the underlying soils. The bactericide applied to plots was ProMac, an anionic surfactant applied in the form of long-term release pellets. Lime requirements to raise the treated soils to pH 7.0 were determined by the SMP agricultural buffer method. The excess lime applied in treatment 5 was determined by acid-base account analysis, simulated weathering tests, and buffer recommendations.

Microbial lipids were extracted from soil and sediment samples using the method of Guckert et al. (1986). The polar lipid fraction (containing the phospholipids) was separated from the other lipid fractions by silicic acid column chromatography. Phospholipid fatty acid methyl esters were prepared by transesterification of the phospholipids. The methyl esters of the fatty acids were then separated and quantified by capillary gas-liquid chromatography and identified by electron impact mass spectrometry (GC/MS). Cluster analysis was accomplished using Ein*Sight (Infometrix, Seattle, Washington) pattern recognition software.

RESULTS

Kwajalein Site

Initial characterization of the Kwajalein site using PLFA analysis indicated a diverse microbial community existed and consisted primarily of actinomycetes (evidenced by tuberculostearic acid) and gram-negative organisms (evidenced by monoenoic PLFA). During the bioremedial activities, SLB analysis showed an overall increase in the mean biomass estimate in the nutrient-amended plots, and the ex situ plots as compared to the controls (Figure 2). Multivariate statistics showed that the control sediments differed slightly from the nutrient-amended plots, and both of these sample groups differed from the ex situ treatments with regards to PLFA composition. These results indicate that the treatments had an effect on the community structure of the extant microbiota. Analysis of individual PLFA indicated that the community which showed the greatest response to nutrient amendments was primarily gram-negative, as evidenced by the increases in the percentages of the monoenoic fatty acids 16:1ω7cis and 18:1ω7cis, terminal points in the anaerobic desaturase pathway utilized by the gram-negative bacteria. Changes in the nutritional status of the gram-negative population were noted as well. Although no change in *trans/cis* ratios (indicator of environmental toxicity) occurred between the control and nutrient-amended plots, the percentage of cyclopropyl 17:0 decreased in the nutrient-amended plots as compared to the controls, indicating some alleviation of nutritional stress during the course of the remediation effort. Although the production of cyclopropyl PLFA has been related to metabolic stress, certain bacterial types have been shown to contain more of this

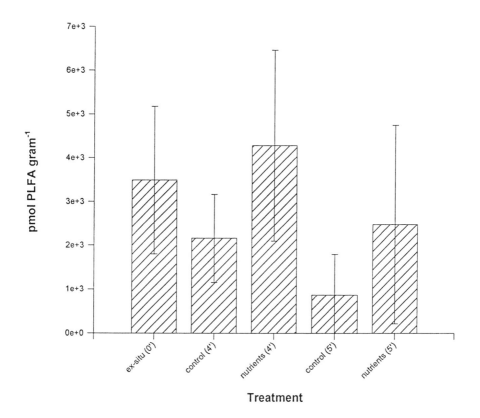

FIGURE 2. Biomass estimations for the Kwajalein Atoll control plots ex situ plots, and nutrient-amended plots.

type of fatty acid than others under ideal growth conditions. Therefore, the observed decrease in this fatty acid could also be related to a shift in community structure.

Montana Mine Waste Site

The investigation of this site was conducted over a 2-year period to determine if the addition of bactericide to these sites could reduce the population of *Thiobacillus ferrooxidans*, the assumed causative agent of soil acidification at this site. Biomarkers specific to the *Thiobacillus* genus were used to determine whether the bactericide was having any effect on the *Thiobacillus* population. The biomarkers used were a cyclopropyl 18:0, a branched cyclopropyl 20:0, and 3-hydroxy-16:0, a hydroxy fatty acid prominent in the lipopolysaccharide (LPS) of *Thiobacillus* species. Values for PLFA recovered from extracted mine wastes are shown in Table 1. Interestingly, it was found that the plot to which no treatments were given showed the lowest overall biomass, and the lowest numbers

TABLE 1. Biomass estimations for the mine waste reclamation site. Biomass is expressed in picomoles PLFA per gram of soil extracted.

Total PLFA (pmol/gram soil)

	Topsoil 1991	Topsoil 1992	Middle soil 1991	Middle soil 1992	Bottom soil 1991	Bottom soil 1992
Plot 1	2.25×10^{15} (± 1.61)	1.26×10^{15} (± 0.06)	2.26×10^{15} (± 1.71)	1.71×10^{15} (± 0.75)	8.47×10^{13} (± 11.5)	1.86×10^{15} (± 1.32)
Plot 2	5.94×10^{15} (± 5.23)	2.17×10^{16} (± 0.67)	5.83×10^{14} (± 4.76)	6.42×10^{15} (± 1.38)	1.41×10^{14} (± 1.25)	2.37×10^{15} (± 1.47)
Plot 3	5.46×10^{16} (± 4.81)	1.91×10^{16} (± 0.42)	9.60×10^{15} (± 1.08)	6.53×10^{15} (± 2.0)	1.43×10^{14} (± 0.64)	2.74×10^{15} (± 1.73)
Plot 4	1.42×10^{16} (± 1.31)	4.71×10^{15} (± 1.16)	3.71×10^{15} (± 1.18)	1.49×10^{15} (± 0.61)	1.02×10^{14} (± 1.01)	2.36×10^{15} (± 1.53)
Plot 5	5.55×10^{16} (± 4.08)	1.98×10^{16} (± 0.40)	9.71×10^{15} (± 2.47)	5.76×10^{15} (± 0.66)	1.65×10^{14} (± 1.15)	3.45×10^{15} (± 0.99)

Thiobacillus ferrooxidans PLFA (pmol/gram soil)

	Topsoil 1991	Topsoil 1992	Middle soil 1991	Middle soil 1992	Bottom soil 1991	Bottom soil 1992
Plot 1	7.48×10^{12} (± 2.08)	2.78×10^{13} (± 2.08)	N/D	1.85×10^{14} (± 1.41)	4.32×10^{12} (± 2.49)	5.96×10^{14} (± 4.15)
Plot 2	8.63×10^{13} (± 3.54)	2.35×10^{14} (± 0.92)	1.91×10^{13} (± 1.66)	2.48×10^{14} (± 0.19)	9.14×10^{12} (± 11.8)	7.31×10^{14} (± 4.62)
Plot 3	3.93×10^{14} (± 3.48)	2.45×10^{14} (± 0.89)	4.33×10^{13} (± 1.47)	2.54×10^{14} (± 0.50)	6.59×10^{12} (± 4.98)	1.12×10^{14} (± 0.84)
Plot 4	N/D	2.29×10^{14} (± 0.98)	2.88×10^{12} (± 1.66)	1.43×10^{14} (± 0.55)	8.31×10^{11} (± 14.4)	9.07×10^{14} (± 6.96)
Plot 5	3.42×10^{14} (± 2.14)	2.28×10^{14} (± 0.97)	3.73×10^{13} (± 1.08)	2.54×10^{14} (± 0.39)	9.14×10^{12} (± 8.01)	1.31×10^{15} (± 0.65)

Thiobacillus genus PLFA (pmol/gram soil)

	Topsoil 1991	Topsoil 1992	Middle soil 991	Middle soil 1992	Bottom soil 1991	Bottom soil 1992
Plot 1	7.35×10^{14} (± 5.55)	3.30×10^{14} (± 0.49)	6.82×10^{14} (± 5.37)	5.88×10^{14} (± 3.73)	7.37×10^{14} (± 6.14)	5.96×10^{14} (± 4.15)
Plot 2	2.53×10^{15} (± 0.77)	1.0×10^{16} (± 0.32)	2.49×10^{15} (± 1.96)	3.08×10^{15} (± 0.73)	1.16×10^{15} (± 1.25)	7.31×10^{14} (± 4.62)
Plot 3	1.62×10^{16} (± 1.40)	9.21×10^{15} (± 2.11)	3.75×10^{15} (± 0.24)	3.07×10^{15} (± 1.07)	1.56×10^{15} (± 1.22)	1.12×10^{14} (± 0.84)
Plot 4	6.97×10^{15} (± 1.55)	2.34×10^{15} (± 0.95)	1.13×10^{15} (± 0.55)	6.14×10^{14} (± 2.50)	9.29×10^{14} (± 6.56)	9.07×10^{14} (± 6.96)
Plot 5	2.35×10^{15} (± 1.76)	$9.15 \times \times10^{15}$ (± 1.85)	3.99×10^{15} (± 1.22)	2.65×10^{15} (± 0.44)	2.06×10^{15} (± 0.48)	1.31×10^{15} (± 0.65)

of *Thiobacillus* as determined by SLB. Treatment 2 (bactericide, lime, and cover soil) showed the highest biomass. The only treatment to show decreases in total biomass and in *Thiobacillus* numbers was treatment 4, i.e., bactericide and lime with no cover soil. Treatments 3 and 5 both showed increases in total microbial biomass and *Thiobacillus* numbers. Cluster analysis showed distinct community structure changes between sample plots indicating that the treatments were, in fact, influencing the microbial communities (Figure 3). All

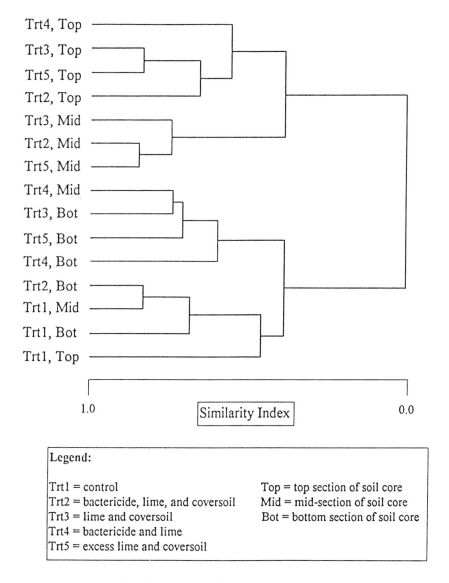

FIGURE 3. Hierarchical cluster analysis of PLFA profiles from treated and untreated soils from the Mine Waste Reclamation Site.

control soils clustered together. The bottom soil of treatment 2 samples also clustered with the control plot, indicating that the treatments applied to this plot had little or no effect at this soil depth. The middle depth soils of all plots treated with a cover soil clustered together, indicating that the cover soil may make a significant contribution to the microbial community structure. All top-soils of the treated plots clustered away from the control plots, suggesting that soil treatments resulted in a change in the microbial community structure. With the exception of treatment 2, the bottom soils of the treated plots clustered away from the control plot, indicating that the treatments could influence the microbial community structure.

DISCUSSION

At the Kwajalein site, SLB analysis indicated an initial microbial community structure containing both actinomycete and gram-negative organisms. These results were supported by other investigators through their use of gene probes and classical microbiological techniques (Adler et al. 1992). The PLFA-based biomass estimations paralleled those of investigators using classical methods, and changes in the nutritional status followed nutrient amendments over the course of the bioremediation effort. By the use of SLB, it was determined that the nutrient amendments successfully increased the extant microbial biomass on site, with an associated shift in the community composition.

At the mine reclamation site, the SLB technique proved very useful for showing the lack of effectiveness of the traditional technique of simply using buffering lime to reclaim acid soil. It also showed that the application of cover soil seemed to enhance both total microbial biomass and that of the *Thiobacillus*. The fact that the control samples contained less biomass suggests a disturbance effect resulting from the stimulation of the microbial populations. These data also suggest that the application of a cover soil not only serves to increase the microbial population, but perhaps diminishes the effect of the lime and bactericide by dilution. This is supported by the clustering of the bottom soils of all the plots away from the treated soils.

The data reported in the peer-reviewed literature clearly demonstrate the efficacy of the SLB in correlating the microbial ecology with bioremediation potential both in the laboratory and the field. This efficacy for prediction between microbial activity and ecology has been demonstrated for TCE and PCE (Phelps et al. 1988, 1990, 1991), PCBs (Hill et al. 1989), linear alkane sulfonate degradation (Federle et al. 1991), and alkanes (Ringelberg et al. 1988). The analysis of other lipids such as the diglycerides (showing nonviable biomass), sterols (for the microeukaryotes—nematodes, algae, protozoa), glycolipids (phototrophs, gram-positive bacteria) (Vestal & White 1989), or the hydroxy fatty acids in the lipid A of lipopolysaccharide (gram-negative bacteria) (Parker et al. 1982) can provide an even more detailed community structure analysis.

ACKNOWLEDGMENTS

The authors wish to acknowledge the support of DOE—Martin Marietta subcontract # 11X-SP830V for its support of the Kwajalein research, and Dr. Douglas J. Dollhopf and the Montana State University College of Agriculture Reclamation Unit for their support of the Montana Mine Waste project. The authors also gratefully acknowledge the technical assistance of Stephen J. Nold.

REFERENCES

Adler, H. I., R. L. Jolley, and T. L. Donaldson (Eds.). 1992. *Bioremediation of petroleum-contaminated soil on Kwajalein Island: Microbiological characterization and biotreatability studies.* Martin Marietta Energy Systems, Inc., ORNL/TM-11925, Oak Ridge National Laboratory, Oak Ridge, TN.

Federle, T. W., R. M. Ventullo, and D. C. White. 1991. "Spatial distribution of microbial biomass, activity, community structure, and xenobiotic biodegradation in the subsurface." In C. B. Fliermans and T. C. Hazen (Eds.), *Proceedings of the First International Symposium on Microbiology of the Deep Subsurface.* Orlando, FL. pp. 7-109 to 7-115.

Guckert, J. B., M. A. Hood, and D. C. White. 1986. "Phospholipid, ester-linked fatty acid profile changes during nutrient deprivation of *Vibrio cholerae:* Increases in the *trans/cis* ratio and proportions of cyclopropyl fatty acids." *Appl. Environ. Microbiol.* 52: 794-801.

Hill, D., T. Phelps, A. Palumbo, D. White, G. Strandberg, and T. Donaldson. 1989. "Bioremediation of polychlorinated biphenyls." *Appl. Biochem. Biotechnol.,* 20/21: 233-243.

Kehrmeyer, S. R., B. M. Appelgate, H. C. Pinkart, D. B. Hedrick, D. C. White, and G. S. Sayler. 1995. "Combined lipid/DNA extraction method for environmental samples." *J. Microbiol. Meth.,* in prep.

Nichols, P. D., J. M. Henson, C. P. Antworth, J. Parsons, J. T. Wilson, and D. C. White. 1987. "Detection of a microbial consortium including type II methanotrophs by use of phospholipid fatty acids in an aerobic halogenated hydrocarbon-degrading soil column enriched with natural gas." *Environ. Toxicol. Chem.* 6: 89-97.

Parker, J. H., G. A. Smith, H. L. Fredrickson, J. R. Vestal and D. C. White. 1982. "A sensitive assay, based on the hydroxy fatty acids from lipopolysaccharide Lipid A for gram-negative bacteria in sediments." *Appl. Environ. Microbiol.* 37: 459-465.

Phelps, T. J., D. H. Ringelberg, D. B. Hedrick, J. D. Davis, C. B. Fliermans, and D. C. White. 1988. "Microbial activities and biomass associated with subsurface environments contaminated with chlorinated hydrocarbons." *Geomicrobiol.* 6: 157-170.

Phelps, T. J., J. J. Niedzielski, R. M. Schram, S. E. Herbes, and D. C. White. 1990. "Biodegradation of trichloroethylene in continuous-recycle expanded-bed bioreactors." *Appl. Environ. Micro.* 56: 1702-1709.

Phelps, T. J., J. J. Niedzielski, K. J. Malachowsky, R. M. Schram, S. E. Herbes, and D. C. White. 1991. "Biodegradation of mixed organic wastes by microbial consortia in continuous-recycle expanded-bed bioreactors." *Environ. Sci. and Tech.* 25: 1461-1465.

Ringelberg, D. B., J. D. Davis, G. A. Smith, S. M. Pfiffner, P. D. Nichols, J. B. Nickels, J. M. Hensen, J. T. Wilson, M. Yates, D. H. Kampbell, H. W. Reed, T. T. Stocksdale, and D. C. White. 1988. "Validation of signature polar lipid fatty acid biomarkers for alkane-utilizing bacteria in soils and subsurface aquifer materials." *F.E.M.S. Microbiol. Ecol.* 62: 39-50.

Shoemaker, H. E., E. O., McLean, and P. F. Pratt. 1961. *Soil Sci. Soc. Amer. Proc.* 25: 274-277.

Vestal, J. R. and D. C. White. 1989. "Lipid analysis in microbial ecology." *Bioscience* 39: 535-541.

Enzyme Assays as Indicators for Biodegradation

Jaap J. van der Waarde, Edwin J. Dijkhuis,
Maurice J.C. Henssen, and Sytze Keuning

ABSTRACT

Laboratory-scale biodegradation studies were used to determine the feasibility of bioremediating sites contaminated with mineral oil. In addition to the CO_2 production, gas chromatograph (GC) measurements, and bacterial numbers, enzyme assays were performed to determine the usefulness of these measurements as indicators of biodegradation. The results indicate that bacterial numbers are indicative of a stimulated biodegradation process, but do not represent an accurate measurement of the actual biodegradation. Dehydrogenase activity, measured with the reduction of triphenyltetrazolium chloride (TTC), appeared to have a good relationship with CO_2 production in several soils. Esterase activity, measured with the hydrolysis of fluorescein diacetate (FDA), was indicative of the onset of biodegradation, but measurement of the decline of activity was less accurate. Application of these parameters to monitor bioremediation in column studies is described in an accompanying paper (van der Waarde et al. 1995).

INTRODUCTION

Full-scale in situ bioremediation is routinely preceded by laboratory studies to determine the feasibility of the technique. Two types of labscale studies are commonly used: batch studies and (undisturbed) column studies. These studies yield information on the biodegradability and bioavailability of the contaminant, limiting factors such as nutrient availability and pH, and maximal contaminant removal rates. To permit a valid comparison between the biodegradation in lab studies and the actual biodegradation process in the field, monitoring instruments are needed that allow a fast and accurate determination of biodegradative activity. An evaluation of enzyme activities as a monitoring instrument for a full-scale bioremediation process was made through comparative research using measurements of dehydrogenase activity and FDA hydrolysis, CO_2 production, bacterial numbers, and GC measurements. The experiments were performed in closed batch experiments, using soil from different sites contaminated with mineral oil.

MATERIALS AND METHODS

Soils

Soils from industrial sites contaminated with mineral oil were used in all studies. If available noncontaminated soil from the same location was used for reference. Contamination levels ranged from 350 to 3,500 mg hydrocarbons/kg dry weight (dw). The soil consisted of sandy aquifer or peat-like soil.

Batch Studies

Soil (45% dw) was incubated in closed batches with and without the addition of inorganic nutrients (KNO_3, Na_2HPO_4, and KH_2PO_4) in a relative concentration of C:N:P = 250:10:5. C was equal to the total contamination level (350 to 3,500 mg/kg dw). Phosphates were added in equal amounts. Flasks were incubated at 20°C under slow rotary shaking. CO_2 evolution was determined using an alkali trap, and the gas phase was regularly replaced to avoid oxygen limitation. Bacterial numbers were determined using the most probable number (MPN) technique, with mineral oil as the sole source of carbon and energy. Hydrocarbons were determined by gas chromatographic analysis preceded by florisyl cleanup to remove humic substances. Dehydrogenase measurements were performed using a modified spectrophotometric method according to Thalmann (1968), and expressed as µg triphenylformazan (TPF) produced per g dw or mL. FDA hydrolysis measurements were performed according to a modified procedure of Schnürer & Rosswall (1982), and expressed as µg fluorescein (F) produced per g dw or mL.

RESULTS

Sandy Aquifer 1, Peat Soil

Both contaminated and noncontaminated soil from a location with both sandy material and peat-like soil was used. The hydrocarbon content in the sandy material decreased from 1,400 to less than 25 mg/kg dw in 28 days. The highest CO_2 production was found in the first week of the incubation (Figure 1). A peak in dehydrogenase activity coincided with the peak in CO_2 production, followed by a mutual decline. Bacterial numbers followed the same pattern. FDA hydrolysis activity still increased after a peak in activity after 7 days. For all measured parameters, these effects were minor or absent in the contaminated soil without nutrients or the noncontaminated soil with nutrients, respectively.

Hydrocarbon content in the peat-like soil decreased from 350 to 280 mg/kg dw in 28 days. The biodegradation process was limited as was shown by a linear

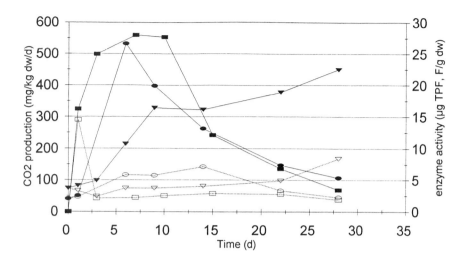

FIGURE 1. Biodegradation of mineral oil in batch incubations (sandy aquifer 1). □ CO_2 production; ○ dehydrogenase activity (μg TPF/g dw); ∇ FDA
hydrolysis (μg F/g dw). Closed symbols = contaminated; open symbols
= noncontaminated.

CO_2 evolution and no decrease in hydrocarbons (Figure 2). Low dehydrogenase
activity could be measured. FDA hydrolysis could not accurately be determined
due to interference of humic substances in the assay. Bacterial numbers did
not increase in the peat, and no effect of nutrient amendment to the peat could
be measured on any of the measured parameters.

Sandy Aquifer 2

The hydrocarbon content in the aerobic incubation decreased from 3,500 to
960 mg/kg dw in 36 days. There was no clear peak in CO_2 production, but
the biological activity reached its maximum in the second week of the incubation
(Figure 3). This maximum coincided with a clear peak in dehydrogenase activity.
After 4 weeks of incubation, CO_2 production and dehydrogenase activity were
still at an elevated level. FDA hydrolysis reached its maximum after 1 week
incubation and did not decline during the measurements. After 2 weeks, no
clear increase in bacterial numbers could be found.

DISCUSSION

Biodegradation of hydrocarbons could be monitored well using CO_2 production and decrease in hydrocarbon content. Effects of low hydrocarbon concentration or the addition of extra inorganic nutrients could directly be found in

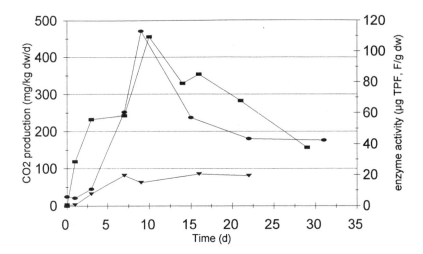

FIGURE 2. Biodegradation of mineral oil in batch incubations (peat soil).
□ CO_2 production; O dehydrogenase activity (µg TPF/g dw); ∇ FDA hydrolysis (µg F/g dw). Closed symbols = contaminated; open symbols = noncontaminated.

a lower or higher CO_2 production respectively. After stimulation of the biodegradation process under optimized conditions, bacterial numbers increased; but this increase was also present in noncontaminated soil or soil without nutrient amendment. Therefore, changes in bacterial numbers are indicative of a stimulated

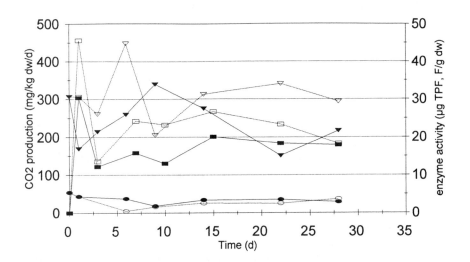

FIGURE 3. Biodegradation of mineral oil in batch incubations (sandy aquifer 2). ■ CO_2 production; ● dehydrogenase activity (µg TPF/g dw); ▼ FDA hydrolysis (µg F/g dw).

biodegradation process, but do not represent an accurate measurement of the actual biodegradation. Dehydrogenase activity, measured with the reduction of TTC, appeared to have a good relationship with CO_2 production in several soils. Biodegradation of contaminant in a peat-like soil did not take place. This process could not be monitored with CO_2 production, bacterial numbers or FDA hydrolysis, but the absence of activity was confirmed with TTC reduction.

FDA hydrolysis was sometimes indicative of the onset of biodegradation, but did not decline after the biodegradation had decreased. Therefore this assay is not a representative measure for actual biodegradation. Furthermore the method is not valid for soils rich in organic carbon, due to interference of humus in the assay.

It was concluded that the parameter that had the best correlation with hydrocarbon removal and CO_2 production was dehydrogenase activity. Further studies are focused on the combined use of this monitoring instrument with standard techniques in undisturbed column studies and pilot and full-scale bioremediation sites.

REFERENCES

Schnürer, J., and T. Rosswall. 1982. "Evaluation of diacetate hydrolysis as a measure of total microbial activity in soil and litter." *Appl. Environ. Microbiol.* 43:1256-1261.

Thalmann, A. 1968. "Zur methodik der bestimmung der dehydrogenaseaktivität im bodem mittels triphenyltetrazoliumchlorid (TTC)." *Landwirtschaftliche Forschung* 21:249-258.

van der Waarde, J. J., E. J. Dijkhuis, S. K. Heijs, M.J.C. Henssen, and S. Keuning. 1995. "Monitoring Bioventing of Soil Contaminated with Mineral Oil." In R. E. Hinchee, R. N. Miller, and P. C. Johnson (Eds.), *In Situ Aeration: Air Sparging, Bioventing, and Related Remediation Processes*. Battelle Press, Columbus, OH. pp. 409-417.

Development of Phylogenetic Probes Specific to *Burkholderia cepacia* G4

Jizhong Zhou and James M. Tiedje

ABSTRACT

Burkholderia (formerly *Pseudomonas*) *cepacia* G4 is capable of oxidizing trichloroethylene (TCE) while growing on toluene or phenol and has potential for use in remediating TCE-contaminated groundwater. To track G4 in the environment, polymerase chain reaction (PCR)-based phylogenetic probes were developed. Four sets of specific primers for PCR amplification of 16S ribosomal RNA gene sequences were designed and tested against both closely and distantly related environmental isolates. Two sets of the primers specifically detected the species *B. cepacia,* and the other two sets specifically detected the strain G4. The sensitivity of these primer sets was evaluated using DNA isolated from the G4 pure culture and from the sterile soils seeded with a known number of G4 and *E. coli* cells. These primer sets were able to detect 1 fg to 1 pg of template DNA from the pure culture, and 2.1×10^2 to 2.1×10^4 G4 cells per g soil in the presence of 1.56×10^8 *E. coli* cells. With such specificity and sensitivity, these probes could be useful in tracking *B. cepacia* G4 and its relatives in the environment.

INTRODUCTION

Trichloroethylene (TCE) is among the most common volatile organic contaminant present in groundwater (Ragagopal 1986). Remediation of TCE-contaminated groundwater using pump-and-treat technology often is inefficient (Abelson 1990; Travis and Doty 1990). In situ bioremediation, in which degradation of pollutants can be enhanced either by stimulation of degradative processes in native microflora or by the addition of pollutant-degrading microorganisms, provides an alternative strategy for the remediation of TCE-contaminated groundwater (Nelson and Bourquin 1990).

Pseudomonas cepacia G4 (Shields et al. 1989), now renamed as *Burkholderia cepacia* G4, and its constitutive mutant strain (Krumme et al. 1993) are capable of efficiently degrading TCE by cooxidation and is considered to be a potential agent for in situ bioremediation. To develop an efficient tool to monitor these organisms in the environment and to evaluate in situ bioremediation of

TCE-contaminated groundwater, the 16S ribosomal RNA gene from *B. cepacia* G4 was sequenced (Zhou et al. 1994b). In this report, gene probes specific to the 16S rRNA genes of species *B. cepacia* and to the strain G4 were developed and tested.

MATERIALS AND METHODS

Organisms and Growth Conditions

The bacterial strains used to evaluate the specificity of the PCR primers were primarily environmental isolates capable of degrading toluene and other aromatic compounds, plus phylogenetically related organisms (Table 1). All of the isolates except *Nitrosomonas europaea*, *Nitrosolobus multiformis*, and *Methylomonas methylovora* were grown for genomic DNA isolation at 30°C on modified R2A (M-R2A) medium as described previously (Fries et al. 1994). The genomic DNA from the nitrifiers and the methylotroph were kindly provided by Mary Ann Bruns of the Center for Microbial Ecology.

Genomic DNA Isolation

Total DNA from bacterial pure cultures was isolated by a SDS-based method (Zhou et al. 1995a). The DNA concentration was measured spectrophotometrically and calibrated with known concentrations of lambda DNA (Pharmacia, Piscataway, New Jersey).

Total DNA from the sterile soils seeded with *B. cepacia* G4 and *Escherichia coli* DHα5F' cells was isolated using a rapid purification method based on gel electrophoresis plus a resin column (Zhou et al. 1995a). G4 and *E. coli* cells in late exponential phase were seeded into sterile soil samples and enumerated by plate count on M-R2A medium. One mL of 10-fold serial dilutions of a G4 cell suspension plus 0.5 mL of *E. coli* cells (3.9×10^{11} cells per mL) were mixed with 5 g of the sterile sandy soil from Wurtsmith Air Base and incubated at room temperature for 1 h prior to DNA extraction. The sterile samples were produced by autoclaving the soil twice at 121°C for 1 h.

PCR Amplification

Species- or strain-specific oligonucleotide primers (Table 2) were designed using the OLIGO program (Rychlick and Rhoads 1989) based on the G4 16S rRNA gene sequence (Zhou et al. 1994b) and other 16S rRNA gene sequences from the Ribosomal Database Project (RDP) (Maidak et al. 1994). The forward primer A and the reverse primers C and D were designed to amplify the 16S rRNA genes of the species *B. cepacia*. The forward primer B coupled with C or D were designed to amplify the 16S rRNA gene from strain G4. When the oligonucleotide primer sequences were run against the RDP SSU database (Release 3.1) by using the RDP CHECK-PROBE service (Maidak et al. 1994), no exact matching complements were found.

TABLE 1. Bacterial strains used for specificity testing.

Strain	Subgroup	BTEX degradation[a]	Accession number[b]	No. of mismatches to primers			
				A	B	C	D
Burkholderia cepacia G4	Beta	BTE	L28675	0	0	0	0
B. cepacia ATCC 25416	Beta	ND	M22518	5	5	1	0
B. pickettii PKO1	Beta	BTE	L37367	5	13	5	3
Alcaligenes eutrophus	Beta	ND	M32021	7	13	5	3
Azoarcus tolulyticus Tol-4[c]	Beta	TE	L33694	5	8	10	5
Nitrosomonas europaea	Beta	ND	M96399	6	12	9	5
Nitrosolobus multiformis	Beta	ND	M96401	7	10	9	5
Methylomonas methylovora	Beta	ND	M95661	8	13	10	4
Pseudomonas putida F1	Gamma	BTE	L37365	8	14	8	4
P. putida PaW1	Gamma	T, *m*, *p*-X	L28676	8	14	8	4
P. mendocina KR	Gamma	TE	L37366	8	13	8	4
CFS 215[d]	Gamma	BTE, *p*-X	Unsubmitted	8	15	8	4
W31[e]	Gamma	BTE	Unsubmitted	9	13	10	4

(a) BTEX (benzene, toluene, ethylbenzene, and xylenes) compounds degraded under aerobic conditions.
(b) 16S rRNA gene sequence accession number in GenBank.
(c) Strain Tol-4 also degrades toluene under anaerobic denitrifying conditions (Chee-Sanford et al. 1992).
(d) Strain CFS 215 also degrades benzene, toluene, ethylbenzene, and *p*-xylene under microaerophilic conditions (Mikesell et al. 1993).
(e) Strain W31 also degrades benzene, toluene, and ethylbenzene under microaerophilic conditions (Mikesell and Olsen 1992).

TABLE 2. Oligonucleotide primers used for PCR amplification.

Primers	Primer sequences	$T_m^{(a)}$ (°C)	*E. coli* 16S rRNA gene positions
A	5'CGCTATAGGGTTGGCCGATGGC3'	61.2	219-240
B	5'CCGAATCTCTTCGGGATTCCGACCA3'	60.0	997-1021
C	5'GGGAGTCTCCTTAGAGTGCTCTTGC3'	60.9	1162-1136
D	5'GCGATTCCAGCTTCATGCACTCGAGT3'	60.8	1344-1319

(a) T_m was calculated by OLIGO program (Rychlick and Rhoads 1989).

The PCR amplification was performed with "hot start" (D'Aquila et al. 1991) in a 20 µL volume containing 1x *Taq* polymerase buffer [10x buffers: 100 mM Tris-Cl, 15 mM $MgCl_2$, 500 mM KCl (pH 8.3)], 200 µM of dNTPs, 2 pmoles of each primer, 1 fg to 100 ng template DNA, and 0.5 unit *Taq* polymerase (Boehringer Mannheim, Indianapolis, Indiana). All stocks for PCR amplification were made and procedures were performed with the precautions suggested by Kwok and Higuchi (1989). The reaction conditions for all primer sets consisted of initial denaturation at 94°C for 2 min, 35 cycles of denaturation at 94°C for 30 s, primer annealing and extension at 74°C for 3 min, plus one additional cycle with a final 6 min of chain elongation. Temperature was cycled in a programmable temperature cycler (GeneAmp PCR System 9600, Perkin Elmer Corp., Norwalk, Connecticut).

Detection of Amplified Products

PCR-amplified DNAs were detected by gel electrophoresis and radiolabeled gene probes. Aliquots of amplified samples (10 µL) were separated by gel electrophoresis on 1.5% agarose gel, stained with ethidium bromide, visualized with a UV transilluminator, and photographed. Southern hybridization was carried out as described previously (Zhou and Tiedje 1995).

RESULTS

Species-Specific Amplification of 16S rDNA Segments

The specificity of the *B. cepacia* species targeted-primer sets (A-C, A-D) was tested with two *B. cepacia* strains and with other members in the beta and gamma subgroups of the *Proteobacteria* listed in Table 1. At an annealing temperature of 74°C, an amplification product of the appropriate size was obtained for all strains of *B. cepacia* on agarose gels. In contrast, no amplification signal was observed for any of the reference strains and the negative control of no DNA (data not shown).

To confirm the specificity by PCR amplification of these primers, the same gels were blotted and hybridized with a 1.1 kb 16S rRNA gene fragment from *B. cepacia* G4. A strong positive signal was detected for both *B. cepacia* species, but no signal was detected for the negative control and all other reference strains except for strains PKO1, F1, and PaW1 (Figure 1a, b). A weaker band was observed for PKO1 with primer set A-C, and a faint band was observed for F1 and PaW1 with primer set A-D, indicating that poor amplification occurred for these strains.

Strain-Specific Amplification of 16S rDNA Segments

The specificity of the strain G4-targeted primer sets (B-C, B-D) was also tested with the same reference strains used above. At the annealing temperature of 74°C, an amplification product corresponding to the expected size of the 16S rRNA genes was observed on agarose gels with strain G4 only and not with the

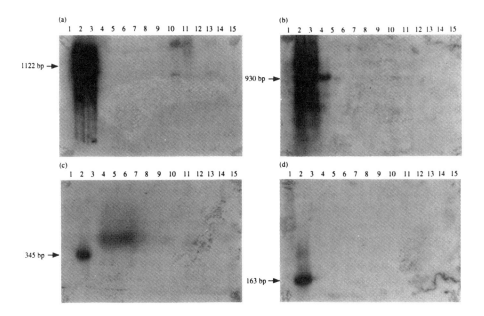

FIGURE 1. Specific detection of *Burkholderia cepacia* G4 by PCR with primer sets A-D (a), A-C (b), B-D (c), and B-C (d). The PCR products were separated on agarose gel, transferred to nylon membrane, and hybridized with a 1.1 kb 16S rRNA gene fragment from *B. cepacia* G4. Lanes: 1, 100 bp ladder size marker; 2, *B. cepacia* G4; 3, *B. cepacia* ATCC 25416; 4, *B. pickettii* PKO1; 5, *Alcaligenes eutrophus*; 6, *Azoarcus tolulyticus* Tol-4; 7, *Nitrosomonas europaea*; 8, *Nitrosolobus multiformis*; 9, *Methylomonas methylovora*; 10, *Pseudomonas putida* F1; 11, *P. putida* PaW1; 12, *P. mendocina* KR; 13, Strain CFS 215; 14, Strain W31; 15, Negative control of no DNA.

other strains (data not shown). No signal was detected by Southern hybridization with all strains except G4 (Figure 1c, d), indicating that these primer sets were specific to strain G4.

Sensitivity Tests with Pure Cultures

To determine the lower limit of detection of G4 DNA, serial 10-fold dilutions of genomic DNA from *B. cepacia* G4 were analyzed by PCR with the four primer sets. On agarose gels, the detection levels were 10 pg for primer sets A-C and A-D, 100 fg for primer set B-C, and 1 pg for primer set B-D (Table 3). The sensitivity of detection was increased at least 10-fold with Southern hybridization (Table 3, Figure 2a for primer set B-D, other not shown).

Sensitivity Tests for G4 Added to Soils

To determine the sensitivity of different primer sets for detecting G4 in heterogeneous DNA background, a known number of G4 cells, ranging from 0 to 1.05×10^{10}, and *E. coli* cells (1.95×10^{11}) were inoculated into 5 g sterile soil; and DNA was then extracted from the soil. Following direct DNA extraction and purification from each seeded soil, a 1 μL aliquot from 1000 μL of purified DNA was used as template for hot-start PCR. Southern hybridization showed that PCR amplification was observed with 2.1×10^2 G4 cells per g soil for primer set B-D (Figure 2b), 2.1×10^2, 2.1×10^3, and 2.1×10^4 G4 cells per g soil for primer sets B-C, A-C, and A-D (data not shown). Assuming a DNA recovery efficiency from soils of 80% (Zhou et al. 1995c), the actual detection limit for PCR amplification by Southern blot was 8 cells for primer set A-D, 8 cells for primer set B-C, 8×10^2 cells for primer set A-C, and 8×10^3 cells for primer set A-D (Table 3).

DISCUSSION

Specificity and sensitivity are the two most important criteria for any assay to become a useful and reliable identification and detection tool to track organisms

TABLE 3. Detection limit of template DNA for different primer sets.

Primer sets	Template DNA from pure culture		Template DNA from seeded soils (cells)	
	Agarose gel	Southern blot	Agarose gel	Southern blot
A-C	10 pg	1 pg	8×10^4	8×10^3
A-D	10 pg	1 pg	8×10^3	8×10^2
B-C	100 fg	1 fg	8×10^1	8
B-D	1 pg	100 fg	8×10^1	8

in environments. In the present study, four sets of PCR primers specific to species *B. cepacia* or to strain G4 were designed and evaluated. Under the annealing temperature of 74°C, specific amplification from the species *B. cepacia* was observed on agarose gels with primer sets A-C and A-D and from strain G4 with primer sets B-C and B-D. These results suggest that these PCR primers appeared to be specific to the species *B. cepacia* or to the strain G4. However, by detecting the PCR products with Southern hybridization, a very faint band corresponding to the size of the 16S rRNA gene was occasionally, but not always, observed with *B. pickettii* PKO1, *P. putida* strains F1, and PaW1 with primer sets A-C and A-D. We tested these strains with these primer sets more than 20 times. Ten to 15% of the time, PCR-amplified 16S rRNA gene products were detected by Southern hybridization for these nontargeted strains. In addition, the specificity of the primer sets appeared to depend on template concentration. With a high concentration of template DNA (e.g., 1 µg genomic DNA in a 20-µL PCR reaction), very poor amplification from some of the reference species also was occasionally observed.

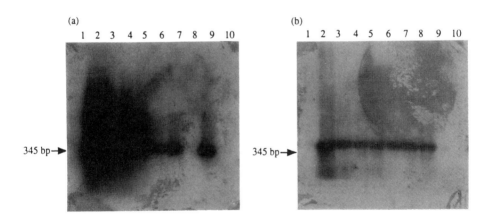

FIGURE 2. Sensitivity of detection by PCR with the primer set B-D on template DNA isolated from a *B. cepacia* G4 pure culture (a) and from the sterile soil seeded with *B. cepacia* G4 and *E. coli* (b). The PCR products were separated on agarose gel, transferred to a nylon membrane, and hybridized with a 1.1 kb 16S rRNA gene fragment from *B. cepacia* G4. (a) Lanes: 1, 100 bp ladder size marker; 2, 10 ng; 3, 1 ng; 4, 100 pg; 5, 10 pg; 6, 1 pg; 7, 100 fg; 8, 10 fg; 9, 1 fg, 10, Negative control of no DNA. (b) Lanes: 1, 100 bp ladder size marker, 2, 2.1×10^9 cells per g soil (8×10^6 cells); 3, 2.1×10^8 cells per g soil (8×10^5 cells); 4, 2.1×10^7 cells per g soil (8×10^4 cells); 5, 2.1×10^6 cells per g soil (8×10^3 cells); 6, 2.1×10^5 cells per g soil (8×10^2 cells); 7, 2.1×10^4 cells per g soil (8×10^1 cells); 8, 2.1×10^3 cells per g soil (8 cells); 9, 10 ng pure *E. coli* DNA; 10. Negative control of no DNA. Lanes 2-8 also were seeded with 3.9×10^{10} *E. coli* cells per g soil (1.56×10^8 cells).

These PCR primers were selected based on the 16S rRNA gene sequences in the current database. Because the database is limited in sequences of these groups, it may not be representative of 16S rRNA sequences of environmental strains. Hence, the specificity of these and often other phylogenetic probes for environmental isolates cannot be fully evaluated. Thus, when using phylogenetic probes to study environmental samples, it should be kept in mind that potential cross amplification or hybridization may occur. Furthermore, other environmental strains may have identical 16S rRNA sequences but differ in TCE-oxidizing ability or other ecological traits. Such strains, however, would be extremely close relatives and hence likely competitors for the same niche. It must also be remembered that phylogenetic probes track the organism and that some transposon- or plasmid-borne traits may be unstable or transfer to other organisms. For strain G4, however, the TCE-oxidizing trait seems to have maintained stable.

With these PCR primers, 1 fg to 1 pg template DNA from a pure culture can be detected, which is equivalent to less than 1 to 200 cells, assuming 5 fg DNA per cell. The sensitivity of PCR amplification with these primers was similar to that achieved in other studies based on 16S rRNA genes (Tsai and Olsen 1992 a, b; Mahony et al. 1993). For primer sets B-C and B-D, the sensitivity of detection using the heterogeneous template DNA (from G4 and *E. coli*) isolated from a soil matrix was comparable to that using the template DNA from the pure culture (Table 3), while for primer sets A-C and A-B, the sensitivity was significantly decreased. The decrease of detection sensitivity could result from insufficient purity of the template DNA or the presence of a large amount of heterogeneous *E. coli* DNA.

The sensitivity of PCR amplification was primer-dependent, and the basis for this is not understood (He et al. 1994). From our study, the sensitivity of primer sets appeared to depend on the size of the amplified fragments, with primers that amplify small fragments giving higher sensitivity. Also, we found that fresh reagents were very important to achieving highly sensitive amplification. In this probe study, we generally used freshly made dNTPs, but in no case more than 10 days old. Furthermore, primer concentration is also important for high-sensitivity PCR amplification. Because primer dimers may occupy the active sites of *Taq* polymerase (Rolfs et al. 1992), the primer concentration should be kept as low as possible. In this study, we used only 1/10 the primer concentration of the standard PCR reaction.

In the first cycles of PCR amplification with genomic DNA as template, primers must perform a "genomic screening" (Rolfs et al. 1992) until they find the complementary annealing sites. The probability of successful primer annealing in the very first cycles is mainly determined by the target copy number and whether there is enough "genomic screening time" to find a target (Rolfs et al. 1992). Rolfs et al. (1992) showed that longer annealing times in the first cycles were advantageous to the genome screening process. With the primer sets A-C and B-C, using a longer annealing time (up to 3 min) increased the sensitivity by 1 to 3 orders of magnitude (data not shown). Thus, we used 3 min for primer annealing and extension for all primer sets to achieve high sensitivity.

Two-step PCR with nested primers is widely used and extremely sensitive (Cassinotti et al. 1993; Zimmermann et al. 1994). When primer sets B-C and B-D were nested with the primer set A-D, however, no increase of the sensitivity of PCR amplification was obtained for primer sets B-C and B-D with the template DNA either from pure culture or from seeded sterile soils. In addition, we also tried "booster PCR" (Ruano et al. 1989; Picard et al. 1992) with the primer sets B-C and B-D, but again the sensitivity was not improved. These results suggested that the increases of sensitivity with nested PCR or booster PCR appear to be primer-dependent.

The specific designed primers were successfully tested in the field. We attempted to track G4 in a bioreactor seeded with G4 for remediation of TCE. Shortly after reactor startup, amplification of the reactor samples was observed with all of these primer sets, but after the reactors operated for 2 months, no amplification was achieved from the samples, indicating that the inoculated strain G4 was replaced by other microorganisms (data not shown). This conclusion was confirmed by the absence of hybridization of reactor DNA to the G4 *ortho*-monooxygenase gene probe (Fries et al. 1994) and the lack of a match of RFLP (restriction fragment length polymorphism) patterns of 16S rRNA genes between the bioreactor community and G4 (data not shown).

With certain precautions, these primer sets can be used to detect G4 and related strains in environmental samples. A diagnostic test based on the combination of these primer sets makes false-positive conclusions less likely but they are not completely excluded because more than one strain can have identical 16S rRNA gene sequences. False-negative results may occur if the number of G4 cells is below the detection limit or if the template DNA from environmental samples is not pure enough. To avoid drawing false-positive or false-negative conclusions, it is advisable to couple these PCR-based phylogenetic probes with some other detection methods, such as DNA or RNA hybridization-based phylogenetic probes (Amann et al. 1990), functional gene probes (Fries et al. 1994), or randomly cloned specific DNA fragments (Thiem et al. 1994). These probes also can be used in the preliminary identification of new isolates, which then can be confirmed by partially sequencing of the hypervariable regions of the 16S rRNA gene, such as the region containing primer B, and other group specific tests. To achieve high specificity and sensitivity, we optimized the conditions for these primer sets on a particular thermocycler (GeneAmp PCR system 9600). The optimal conditions for these primer sets could be slightly different for other thermocyclers.

ACKNOWLEDGMENTS

This project was supported by the Center for Microbial Ecology by the National Science Foundation, Grant No. BIR9120006, and by the National Institute of Environmental Health Sciences Superfund Research and Education Grant No. ES-04911. We thank Mary Ann Bruns for providing genomic DNA and Marcos Fries for collaboration on the bioreactor study.

REFERENCES

Abelson, P. H. 1990. "Inefficient remediation of groundwater pollution." *Science.* 250: 733.

Amann, R. I., L. Krumholz, and D. A. Stahl. 1990. "Fluorescent oligonucleotide probing of whole cells for determinative, phylogenetic and environmental studies in microbiology." *J. Bacteriol.* 172: 762-770.

Cassinotti, P., M. Weitz, and G. Siegl. 1993. "Human parvovirus B19 infections: routine diagnosis by a new nested polymerase chain reaction assay." *J. Med. Virol.* 40: 228-234.

Chee-Sanford, J., M. R. Fries, and J. M. Tiedje. 1992. "Anaerobic degradation of toluene under denitrifying conditions in bacterial isolate Tol-4," Abst. Q233, p. 374. *Abstr. 92th Annu. Meet. Am. Soc. Microbiol. 1992.*

D'Aquila, R. T., L .J. Bechtel, J. A. Videler, J. J. Eron, P. Gorczca, and J. C. Kaplan. 1991. "Maximizing sensitivity and specificity of PCR by preamplification heating." *Nucleic Acids Res.* 19: 3749.

Fries, M. R., J.-Z. Zhou, J. Chee-Sanford, and J. M. Tiedje. 1994. "Isolation, characterization and distribution of denitrifying toluene degraders from a variety of habitats." *Appl. Environ. Microbiol.* 60: 2802-2810.

He, Q., M. Marjamaki, H. Soini, J. Mertsola, and M. K. Viljanen. 1994. "Primers are decisive for sensitivity of PCR." *BioTechniques.* 17: 82-87.

Krumme, M. L., K. N. Timmis, and D. F. Dwyer. 1993. "Degradation of trichloroethylene by *Pseudomonas cepacia* G4 and the constitutive mutant strain G4 5223 PR1 in aquifer microcosms." *Appl. Environ. Microbiol.* 59: 2746-2749.

Kwok, S., and R. Higuchi. 1989. "Avoiding positives with PCR." *Nature.* 339: 237-238.

Mahony, J. B., K. E. Lutinstra, J. W. Sellors, and M. A. Chernesky. 1993. "Comparison of plasmid- and chromosome-based polymerase chain reaction assays for detecting *Chlamydia trachomatis* nucleic acids." *J. Clin. Microbiol.* 31: 1753-1758.

Maidak, B. L., N. Larsen, M. J. McCaughey, R. Overbeek, G. J. Olsen, K. Fogel, J. Blandy, and C. R. Woese. 1994. "The ribosomal database project." *Nucleic Acids Res.* 22: 3485-3487.

Mikesell, M. D., and R. H. Olsen. 1992. "Degradation of aromatic hydrocarbons under anoxic conditions by *Pseudomonas* sp. W31," Abst. Q109, p. 353. *Abstr. 92th Annu. Meet. Am. Soc. Microbiol. 1992.*

Mikesell, M. D., J. J. Kukor, and R. H. Olsen. 1993. "Metabolic diversity of aromatic hydrocarbon-degrading bacteria from a petroleum-contaminated aquifer." *Biodegradation.* 4: 155-164.

Nelson, M.J.K., and A. W. Boruquin. May 1990. "Methods for stimulating biodegradation of halogenated aliphatic hydrocarbons." U.S. patent 4,925,802.

Picard, C., C. Ponsonnet, E. Paget, X. Nesme, and P. Simonet. 1992. "Detection and enumeration of bacteria in soil by direct DNA extraction and polymerase chain reaction." *Appl. Environ. Microbiol.* 58: 2717-2722.

Ragagopal, R. 1986. "Conceptual design for a groundwater quality monitoring strategy." *Environ. Prof.* 8: 244-264.

Rolfs, A., I. Schuller, U. Finckh, and I. Weber-Rolfs. 1992. *PCR: Clinical diagnostics and research.* Springer-Verlag, New York, NY.

Ruano, G., W. Fenton, and K. K. Kidd. 1989. "Biphastic amplification of very dilute DNA samples via "booster" PCR." *Nucleic Acids Res.* 17: 5407.

Rychlick, W., and R. Rhoads. 1989. "A computer program for choosing optimal oligonucleotides for filter hybridization, sequencing and *in vitro* amplification of DNA." *Nucleic Acids Res.* 17: 8543-8551.

Shields, M. S., S. O. Montgomery, P. J. Chapman, S. M. Cuskey, and P. H. Pritchard. 1989. "Novel pathway of toluene catabolism in the trichloroethylene-degrading bacterium G4." *Appl. Environ. Microbiol.* 55: 1624-1629.

Thiem, S. M., M. L. Krumme, R. L. Smith, and J. M. Tiedje. 1994. "Use of molecular techniques to evaluate the survival of a microorganism injected into an aquifer." *Appl. Environ. Microbiol. 60*: 1059-1067.

Travis, C. C., and C. B. Doty. 1990. "Can contaminated aquifers at Superfund sites be remediated?" *Environ. Sci. Technol. 24*: 1464-1466.

Tsai, Y.-L., and B. H. Olsen. 1992a. "Detection of low copy numbers of bacterial cells in soils and sediments by polymerase chain reaction." *Appl. Environ. Microbiol. 58*: 754-757.

Tsai, Y.-L., and B. H. Olsen. 1992b. "Rapid method for separation of bacterial DNA from humic substances in sediments for polymerase chain reaction." *Appl. Environ. Microbiol. 58*: 2292-2295.

Zhou, J.-Z., J. Chee-Sanford, R. Weller, and J. M. Tiedje. 1995b. "The phylogeny of five well-studied, aerobic, toluene-degrading bacteria." Unpublished.

Zhou, J.-Z., M. R. Fries, J. Chee-Sanford, and J. M. Tiedje. 1995a. "Phylogenetic analysis of a new group of denitrifiers capable of anaerobic growth on toluene: Description of *Azoarcus tolulyticus* sp. nov." *Int. J. Syst. Bacteriol.* In press.

Zhou, J.-Z., M. A. Bruns, and J. M. Tiedje. 1995c. "Rapid method for the recovery of DNA from soils of diverse composition." *Appl. Environ. Microbiol*, submitted.

Zhou, J.-Z., and J. M. Tiedje. 1995. "Gene transfer from a bacterium injected into an aquifer to an indigenous bacterium." *Mol. Ecol.*, submitted.

Zimmermann, K., K. Pischinger, and J. W. Mannhalter. 1994. "Nested primer PCR detection of HIV-1 in the background of increasing number of lysed cells." *BioTechniques. 17*: 18-20.

Modeling Shoreline Bioremediation: Continuous Flow and Seawater Exchange Columns

Svein Ramstad, Per Sveum,
Cathe Bech, and Liv-Guri Faksness

ABSTRACT

This paper describes the design and use of the columns in the study of bioremediation processes, and gives some results from an experiment designed to study the effects of different additives (fish meal, stick water, and Max Bac) on biodegradation of crude oil. There is significant difference in oil degradation (nC_{17}/pristane ratio) between the column with additives and those without. Open system models in this type of open column give valuable data on how the chemical and biological parameters, including oil degradation, are affected by the additives, and simultaneously by the dilutive effect of seawater washing through the sediment, and for optimizing formulations. The system is designed with a large number of units and provides a good first approximation for mesocosm studies and field experiments, thus reducing the need for large numbers of such resource-demanding experiments.

INTRODUCTION

Due to environmental regulations, permits to do any field experiment that involves releases into the environment, to study fate, behavior, effects, or treatment of oil on shorelines, can be difficult to obtain. Furthermore, field experiments are expensive and can be difficult to carry out in a controlled manner. Thus, experimental shoreline models in the laboratory can be vital for future studies of oil biodegradation and bioremediation of shorelines. In any type of model studies, whether physical or numerical, it is imperative that the studied parameters are of biological importance and, thus, that the results are relevant to natural systems. Several attempts to simulate beach systems have been reported earlier, either in the form of columns (Johnston 1970; Gibbs and Davis 1982) used for oil spill studies or as more complex beach systems (Pugh 1975) that were not used for oil spill studies.

Shoreline processes are very complex, and the experimental or test systems should at least simulate the main environmental processes, thereby giving the testing validity and a reasonable probability that the findings can be transferred to the real world. At the same time, the experimental system should be simple enough to be reproduced and to allow multiple tests to be conducted simultaneously. It is important that physical/chemical and biological processes can be studied together, as they are strongly correlated. Shoreline systems are characterized by several dynamic processes. Water filtering through the sediment is significant and is likely to be critical in determining the rate of oil biodegradation. The amount of water permeating through the sediment depends on the sediment characteristics, the beach topography, and the tide, and can exceed more than $18 \text{ m}^3 \cdot \text{m}^{-1} \cdot \text{day}^{-1}$ (McLachlan 1982; Riedl 1971). It affects biodegradation of oil by simultaneously enhancing the physical self-cleaning of the sediment (Sveum and Bech 1994), and by counteracting fertilizer action by removing added fertilizers and bacteria. These complex processes can be simulated only in an open system with continuous flow and exchange of seawater.

Two types of shoreline models for shoreline biodegradation and bioremediation studies have been developed; continuous flow basins and continuous flow columns. Both systems are open, with the desirable exchange of seawater as the main feature. The continuous flow basins are described by Sveum et al. (1985, this volume). The shoreline models have been used both in testing shoreline bioremediation additives and for general biodegradation studies.

The two different experimental systems have been designed to allow for factorial statistical treatment of the experimental results, as 16 (2^4) columns and 4 (2^2) basins are in operation. The experimental systems are operated under standardized conditions, which allows comparison of results between different experimental series. The column system is designed and well suited for multifactor experiments with limited resource use. Because the sediment volume and the volume of oil and additives used can be controlled, mass balance studies can be performed. This paper describes the design and use of the columns in the study of bioremediation processes and gives some results from an experiment designed to study the effects of different additives on biodegradation of crude oil.

PHYSICAL DESCRIPTION AND OPERATION OF CONTINUOUS FLOW SEAWATER EXCHANGE COLUMNS

The column system is operated in a constant-temperature room, with no exposure to light. Fresh seawater is pumped from the sea, and the water is discarded after use without any recycling. The standard temperature of the seawater varies between 10 and 12°C and is only slightly dependent on seasonal variations.

The columns are made from Plexiglas™ pipes (10 cm inner diameter and a total length of 70 cm) (Figure 1) and are equipped with sampling ports located

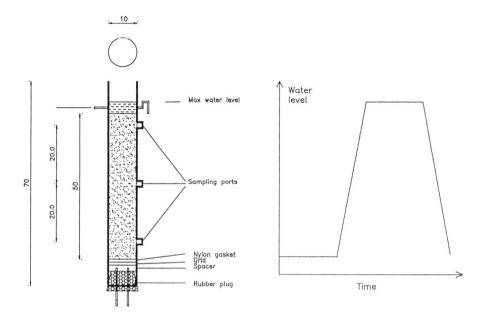

FIGURE 1. Design of continuous flow column and variation of the water level in the column during one tidal cycle.

20, 35, and 50 cm from the top of the column. The sampling ports are closed with rubber stoppers.

The column is closed in the bottom with a rubber stopper. The sediment in the column rests on a stainless steel grid and a nylon gasket situated on a spacer 20 mm above the rubber stopper. Water is supplied and drained through the bottom of the column, and the water flow through the column at high tide is removed through the overflow system. Overflow water is drained through a pipe located 12 cm above the top of the sediment at high tide, which is designed as a drain trap with a small hole at the top that allows accurate (±2 mm) control of the water level. A water sampling pipe is located 15 cm below the top of the column for monitoring chemical and physical parameters in the water. The water flow in the column simulates tidal variation and water circulation in the sediment. The simulation of tidal variation is regulated with a variable-speed peristaltic pump at the water inlet tube and a magnetic valve at the drainage tube (Figure 2). Water is supplied by the pump from the seawater reservoir to the column system at rising and high tides with the magnetic valve closed. At falling and low tides, the magnetic valve is open with no supply of water. The column drainage rate is controlled by natural gravitation and the diameter and length of the drainage tubing. The water level in the columns is monitored with a pressure sensor at the bottom of one of the columns in each of the eight-column sets.

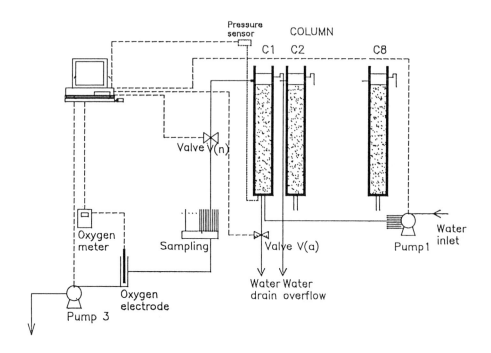

FIGURE 2. Control system for water flow and monitoring of physical parameters in the column equipment.

With minor modifications of the system, other modes of water flow can be designed; e.g., flow from top to bottom or supply at the top and removal through the normal overflow system.

The control program "CONFLOW"(Control System for Continuous Flow Column) is operated under Lab Windows (National Instruments) on an 486 personal computer. The program which operates the two sets of eight columns simultaneously with a multifunctional I/O-board and a digital I/O board has options for configuration and start of experiments through graphical user interface; control of pumps, valves, and measuring devices; modification of experimental setup during runs; monitoring, logging, and control of oxygen concentrations and water pressure; and graphing of logged parameters and options for monitoring five additional parameters in the column itself or in the flow cell.

The water oxygen concentration is measured in water sampled from the top of the column during high tide using a peristaltic pump and a flow cell with a polarographic oxygen electrode. Samples are pumped from each of the columns through a manifold with one magnetic valve for every column and a valve for fresh seawater. The fresh seawater is pumped through the flow chamber to equilibrate the electrode. Measurement of the oxygen content in the column seawater starts 30 min after high tide conditions have been established. Each column is analyzed at 5-min intervals by averaging 10 individual readings obtained during a 4-min period.

Water and sediment samples can be collected during and at the end of the experiments. Sediment can be sampled from both the sediment surface and from the sampling ports. At the termination of the experiment, the entire sediment can be removed from the column by a hydraulic system that allows sampling of sediment layers of defined thickness. Water samples can be taken from the top of the column or through the sampling ports at high tide. Experimental design parameters can be varied to meet the experimental objectives, for example, sediment height, type, and origin (e.g., intact core samples); salinity, temperature, and nutrient levels of inflowing water; water flowrate and tide parameters; and ambient temperature and light exposure.

The columns usually are filled with sediment to a height of 50 cm. The system is normally operated without oil or additives for 1 week under normal conditions to establish marine conditions in the column. The tidal cycles are 240, 120, 240, and 120 minutes for low, rising, high, and falling tide, respectively. The overflow rate at high tide is 0.75 L/h. Only naturally occurring bacteria present in the seawater and the sediment are used. The oil is applied at the water surface during high tide and forms a film of equal thickness, which gives a well-distributed oil phase in the upper strata of the sediment. Additives are supplied to the sediment according to the experimental objectives during the low tide.

Between each experiment, the individual parts of the column system are washed and maintained.

MATERIALS AND METHODS

To describe the use of the column system, results from some experimental series are presented. These experiments were performed under standardized conditions (see above) with gravel in the size range of 0 to 40 mm in the column. Topped Statfjord crude oil (150+) (0.96 kg $\cdot$ m^{-2}) was used as the hydrocarbon source. The additives were supplied in a concentration of 2.8 g fish meal and stick water (Sveum et al. 1994) and 1.4 g Max Bac (Grace Sierra).

Microbial activity (esterase activity) in the sediment samples was analyzed according to a modified procedure of Schnürer and Rosswall (1982). Samples of sediment (approximately 1 g) were incubated with TRIS buffer (60 mM in filtered seawater, pH 8.1, 10 mL) and FDA solution (250 mL, 2 g $\cdot$ L^{-1} in acetone) in sterile test tubes for 60 min at room temperature on a rotary mixer (15 rpm). The samples were filtered through a coarse filter (black band) and a sterile filter (0.2 μm) before measurement.

The concentration of nutrients in interstitial water was determined colorimetrically with an automated system for nutrient analysis (Aquatec, Tecator AB). The total concentration of nitrogen in water samples was measured as nitrite after digestion and oxidation. The water samples were analyzed directly for both nitrate and ammonia.

Hydrocarbons in the sediment samples (3 to 15 g) were extracted in a soxhlet extraction system (Soxtec System, Tecator) with hexane (40 mL) at 140°C for 2 h.

The samples were analyzed with gas chromatography/flame ionization detector (GC/FID) (Hewlett-Packard Series II Gas Chromatograph with splitless injector) after filtration (0.2 μm) and dilution to the appropriate concentration.

RESULTS AND DISCUSSION

A change in the chemical composition of the oil, measured as the nC_{17}/pristane ratio, was observed in all columns (Figure 3a). The most pronounced changes were found in the column with fish meal and stick water. The nC_{17}/pristane ratio also decreased in the control column with no additives, but not as much as in those with additives. The columns with Max Bac and stick water were significantly different from the control (Tukey-Kramer All Pair Test, Abs(Dif)-LSD: 0.05 and 0.12; cf. SAS 1989 for statistical details).

The addition of organic additives resulted in a high initial increase in the esterase activity during the first week (Figure 3b), followed by a moderate increase with stick water and a more pronounced increase in the second half of the experiment. In the column with Max Bac, the esterase activity increased insignificantly during the first week and remained at the same level until the

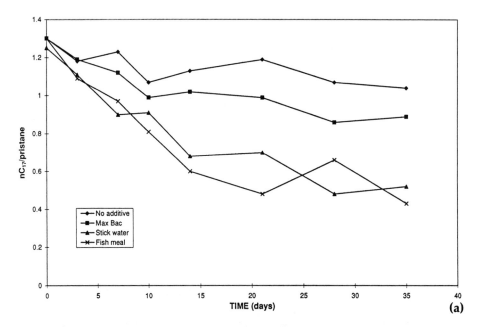

FIGURE 3. Change in (a) nC_{17}/pristane ratio in a series of column experiments with one inorganic additive (Max Bac), two organic additives (fish meal and stick water) and a reference column without additive. All samples were taken at the top of the sediment.

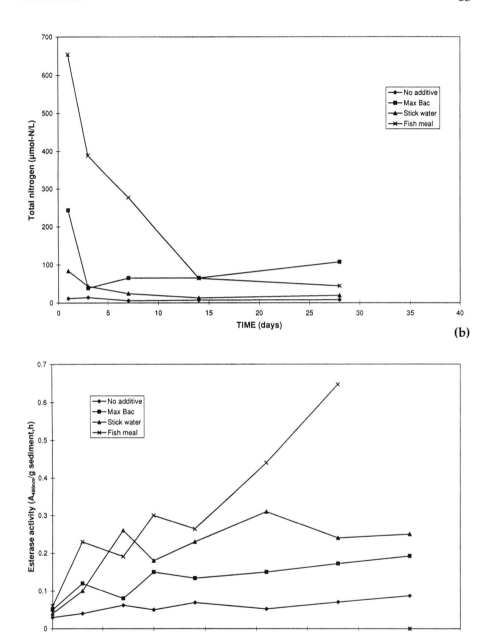

FIGURE 3 (continued). Change in (b) esterase activity, and (c) total soluble nitrogen in a series of column experiments with one inorganic additive (Max Bac), two organic additives (fish meal and stick water) and a reference column without additive. All samples were taken at the top of the sediment.

end of the experiment. The esterase activity in the sediment without additive was much lower than with additives and significantly lower than in the column with stick water (Tukey-Kramer All pair test, Abs(Dif)-LSD: 0.002).

The data from the different experiments show that the concentration of nitrogen varies with the different additives (Figure 3c). In the column with fish meal, there was a initial high concentration of nitrogen, which decreased during the experiment. With stick water, the nitrogen concentration also decreased, but less than with fish meal. With Max Bac, the initial concentration was high followed by a rapid decrease at day 3, probably due to release of nutrients from grains with defect surfaces. Later, the nitrogen concentration increased, due to the natural release from Max Bac. Without additives the concentration was low and approximately constant during the entire experiment.

The results given above show how the different additives affect some chemical and biological parameters in the columns. As expected, the experimental results clearly show that there is significant difference between the column with additives and those without. Open system models in this type of open column give valuable data on how the chemical and biological parameters, including oil degradation, are affected by the additives and simultaneously by the dilutive effect of seawater washing through the sediment.

In addition to the parameters studied in this experiment, a large number of other microbial and chemical parameters can be included. By increasing the number of analyses and sampling points, the mechanism and kinetics of the oil degradation process with the different additives could be documented more easily. As shown by the results above, there was some variation from one sampling point to another, mainly due to the nature of the sediment. The surface of the sediment in the column is restricted, and a limited number of samples can therefore be taken during one experiment. In addition, it is possible that wall effects can affect the results.

Another strategy for the column experiments is to sample at the termination of the experiment. The intact sediment is removed by use of the hydraulic system as described above. This will reduce sampling problems. As these layered samples are larger, analysis can be replicated. Problems with heterogeneous distribution of the oil and the nutrient, as well as sediment heterogeneity, can be avoided. A total mass balance for both the oil and the added nutrient can be established. Information on the subsurface degradation of the oil components can also be studied. If this type of sampling is combined with replication of column combinations and removal of individual columns over time, the column system could be used to study the time dependency of the biodegradation processes with a high degree of confidence.

The column system has been used continuously for more than 2 years without technical problems. Experiments done under standardized conditions allow comparison of different experiments. The results obtained in different experiments under the same conditions showed that the results obtained could be reproduced and the deviation in results could be associated primarily with the variation in the sediment properties. With organic additives, the main problem has been development of anaerobic conditions in the sediment. This problem can be reduced by increasing the flowrate through the column, which is the only source of energy input to the system. The continuous flow and exchange columns are constructed as a flexible system to allow for modification of the columns themselves, the water flow patterns, and other experimental parameters.

CONCLUSIONS

The column system with continuous flow and exchange of seawater is a suitable apparatus for the physical simulation of biodegradation and bioremediation processes in an open system in the laboratory. The system simulates a number of environmental parameters that will affect the degradation processes of petroleum products and takes into account both the chemical and physical properties of the additives. The column system can be used to study the mechanisms and kinetics of the biodegradation processes, for screening and comparing the effect of different additives, and for optimizing formulations. The system is designed with a large number of units, and provides a good first approximation for mesocosm studies and field experiments, reducing the need for large numbers of resource-consuming experiments.

ACKNOWLEDGMENTS

This study is part of the Esso SINTEF Coastal Oil Treatment Program (ESCOST), which is supported by Esso Norge A/S. Special thanks are due to Dr. Roger Prince, Exxon Research and Engineering, and Mr. Geir Indrebø, Esso Norge A/S, for their continuous advice and support throughout the ESCOST program.

REFERENCES

Boucher, G. and S. Chamroux. 1976. "Bacterial and meiofauna in experimental sand ecosystems. I. Materials and preliminary results." *J. Exp. Mar. Biol. Ecol.* 24: 237-249.

Gibbs, G. F. and S. J. Davis. 1982. "The rate of microbial degradation of oil in a beach gravel column." *Microb. Ecol.* 3: 55.

Johnston, R. 1970. "The decomposition of crude oil residue in sand columns." *J. Mar. Biol. Assoc., U.K.,* 50: 925.

McLachlan, A. 1982. "A model for estimation of water filtration and nutrient generation by exposed sandy beaches." *Marine Environmental Research* 6: 37-47.

Pugh, K. B. 1975. "A model beach system." *J. Exp. Mar. Biol. Ecol.* 18: 197.

Riedl, R. J. 1971. "How much water passes through sandy beaches?" *Int. Revue Res. Hydrobiol.* 56: 923-946.

SAS, 1989. *JMP® User's Guide.* Version 2 of JMP.

Schnürer, J. and T. Rosswall. 1982. "Fluorescein diacetate hydrolysis as a measure of total microbial activity in soil and litter." *Appl. Environ. Microbiol.* 43: 1256-1261.

Sveum, P. and C. Bech. 1994. "Bioremediation and physical removal of oil on shore." In R. E. Hinchee, B. C. Alleman, R. E. Hoeppel, and R. N. Miller (Eds.), *Hydrocarbon Bioremediation*, pp. 311-317. Lewis Publishers, Boca Raton, FL.

Sveum, P., L. G. Faksness, and S. Ramstad. 1994. "Bioremediation of oil-contaminated shorelines: The role of carbon in fertilizers." In R. E. Hinchee, B. C. Alleman, R. E. Hoeppel, and R. N. Miller (Eds.), *Hydrocarbon Remediation*, pp. 163-174. Lewis Publishers, Boca Raton, FL.

Sveum, P., S. Ramstad, L. G. Faksness, C. Bech, and B. Johansen. 1995. "Physical modeling of shoreline bioremediation: Continuous flow mesoscale basins." In R. E. Hinchee, G. S. Douglas, and S. K. Ong (Eds.), *Monitoring and Verification of Bioremediation*, pp. 87-96. Battelle Press, Columbus, OH.

Physical Modeling of Shoreline Bioremediation: Continuous Flow Mesoscale Basins

Per Sveum, Svein Ramstad, Liv-Guri Faksness,
Cathe Bech, and Bror Johansen

ABSTRACT

This paper describes the design and use of continuous flow basin beach models in the study of bioremediation processes, and gives some results from an experiment designed to study the effects of different strategies for adding fertilizers. The continuous flow experimental basin system simulates an open system with natural tidal variation, wave action, and continuous supply and exchange of seawater. Biodegradation and bioremediation processes can thus be tested close to natural conditions. Results obtained using the models show a significant enhancement of biodegradation of oil in a sediment treated with an organic nutrient source, increased nutrient level in the interstitial water, and sediment microbial activity. These physical models gives biologically significant results, and can be used to simulate biodegradation and bioremediation in natural systems.

INTRODUCTION

Due to environmental regulations, permits for field experiments to study the fate, behavior, effects, or treatment of oil on shorelines can be difficult to obtain. Furthermore, field experiments are expensive and can be difficult to carry out in a controlled manner. Thus experimental shoreline models in the laboratory can be important for future studies of oil biodegradation and bioremediation on shorelines. In any type of model studies, either physical or numerical, it is vital that the studied parameters are of biological importance, and thus, that the results are relevant to natural systems. Several attempts to simulate beach systems have been reported earlier, either in the form of columns (Johnston 1970; Gibbs and Davis 1982), used for oil spill studies, or as more complex beach systems (Boucher and Chamroux 1976; Pugh 1975), which were not used for oil spill studies.

Shoreline processes are very complex, and the experimental or test systems should at least simulate the main environmental processes. Only this may give the testing validity and a reasonable probability that the findings can be transferred to the real world. At the same time, the experimental system should be simple enough to be reproducible and to allow multiple tests to be conducted simultaneously. It is important that physical/chemical and biological processes be studied together, as they are strongly correlated. Shoreline systems are characterized by several dynamic processes. A very important one is water filtering through the sediment, which is likely to be important in determining the rate of oil biodegradation. The amount of water permeating the sediment depends on the sediment characteristics, as well as on the beach topography and tide, and can exceed more than $18 \, m^3 \cdot m^{-1} \cdot day^{-1}$ (McLachlan 1982; Riedl 1971). It affects biodegradation of oil by enhancing the physical self-cleaning of the sediment (Sveum and Bech 1994) and, simultaneously, by counteracting fertilizer action by removing added fertilizers and bacteria. These complex processes can only be simulated in an open system with continuous exchange of seawater.

We have developed two types of shoreline models for biodegradation and bioremediation studies; continuous flow basins and continuous flow columns. Both systems are open, with the desirable exchange of seawater as the main feature. The continuous flow columns are described by Ramstad et al. (1995, this volume). The shoreline models have been used both for testing shoreline bioremediation additives, and for general biodegradation studies.

The different experimental systems are designed to allow for factorial statistical treatment of the experimental results, as 16 (2^4) columns and 4 (2^2) basins are in operation. The experimental systems are operated under standardized conditions, which allow comparison of results between different experimental series. The column system is designed and well suited for multifactor experiments with limited resource use. Due to the sediment volume and the volume of oil and additives used, mass balance studies can be performed. The main advantages of the basin system over the columns are their size, the possibility of simulating environmental gradients related to beach slope and tidal effects, increased sample size, and the possibility of replicate sampling. The beach models are the last step before field experiments.

This paper describes the design and use of the beach models in the study of bioremediation processes and gives some results from an experiment designed to study the effects of different strategies for adding fertilizer nutrients.

PHYSICAL DESCRIPTION AND OPERATION OF CONTINUOUS FLOW BASINS

Hardware

The continuous flow experimental basin system simulates an open system with natural tidal variation, wave action, and continuous supply and exchange of seawater. Biodegradation and bioremediation processes can thus be tested close to natural conditions.

The design of the basins is shown schematically in Figure 1a. Each basin is 4 × 2 × 1 m; length, width and height, respectively. The basin and the piping are made of polyethylene (PE). The basin is constructed of 10-mm PE plates, supported within an aluminum frame. The pipelines for supply and removal of seawater are constructed of 50-mm PE pipes.

Seawater is supplied to the lower end of the basin through a 5-cm PE pipe, with 3-mm holes every 3 cm, across the total width of the basin. The water outlet is located at the bottom of the basin, 1 m from the lower end and in the middle of the basin. A water overflow safety system is also included. This is designed as a drain lock system with the outflow near the bottom of the basin and piping 95 cm above the bottom of the basin (Figure 1b).

The basin is filled with sediment as shown in Figure 1a, and the shoreline is given a constant slope from the top to the bottom. Standard beach dimensions are 300 cm in length and 90 cm in height.

A wave generator is at the lower end of the basin to provide energy input to the system, thereby avoiding anaerobic conditions in the sediment (Figure 1c). Anaerobic conditions have been observed in experiments without wave energy input. The wave generator flap (2 × 1 m × 1 cm PE) is fastened by a hinge to the bottom of the basin, 10 cm from the wall of the basin. The wave generator blades have 5-cm PE baffles the full height of the plate mounted vertically every 25 cm. To reduce the water pressure behind the plate at high water levels, 5-cm-diameter holes are made. At the top, the plate is attached to an electrical motor (SEW motor, R32DT 60 rpm), which can be regulated continuously (0 to 60 strokes per min) with a speed controller (Scandilogo SLP-750-1 frequency transformer), giving a wave amplitude between 5 and 10 cm.

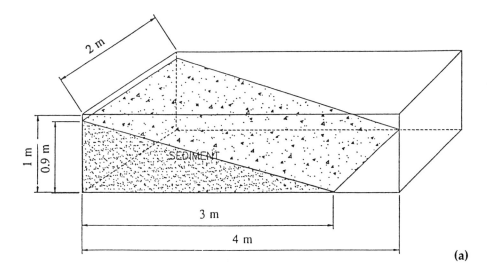

FIGURE 1. Design of continuous flow basins. (a) Physical design of the mesoscale basin.

Control System

The computer-based control and regulation system for the tide is shown in Figure 1b. The water level is controlled by pressure sensors that are interfaced to the computer and is regulated with a magnetic valve at the water inlet. The valves are automatically closed on defined errors on the peripheral hardware system. The water outlet pipe is always open. The flowrate out of the basin is dependent on the water level in the basin and, therefore, varies during the tidal cycle.

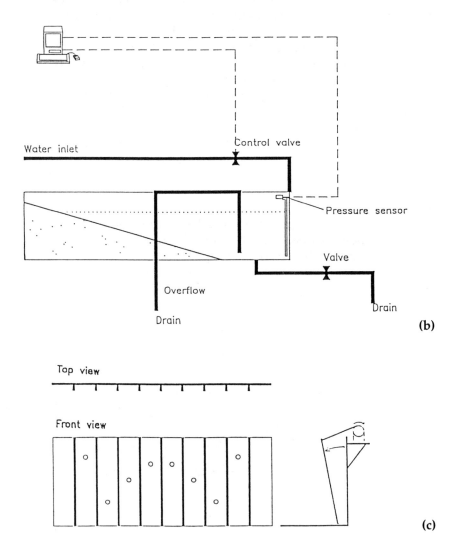

(b)

(c)

FIGURE 1 (continued). Design of continuous flow basins. (b) Water flow and control system. (c) Wave generator.

The control program, "TIDAL SIMULATION," is operated under Lab View (National Instruments) on an Apple Macintosh computer. The program, which controls four basins simultaneously, has options for phase displacement of the tidal cycle, storage of data for documentation, individual calibration of the pressure sensors, graphical display of the water level for the last 24 h, and variation of the amplitude of the tide.

Standardized Experimental Procedure and Basin Operation

The basin experiments are operated with a tidal cycle of 12 h, with the tidal level following a sinusoidal curve. The wave generator is operated at a speed of 15 strokes per min and an amplitude of 7 cm. The wave amplitude at the water surface varies during the tidal cycle, due to decreasing effective stroke length with falling tide. Because the incoming seawater originates at a depth of 70 m, some 1,200 m from the shore, the water temperature varies between 10 and 12°C and is independent of seasonal variations. Ambient air temperature in the laboratory varies between 16 to 18°C. The average seawater exchange rate is 1,200 L/h for each basin. Only bacteria indigenous to the seawater and the sediment were used in these basin experiments.

The experiment is initiated with the application of oil at high tide, without waves. Oil is released to form a film of equal thickness above the intertidal section of the shoreline. The oil is contained in position by a boom during the falling tide (Figure 2a).

Sediment and Water Sampling

Sediment samples from the intertidal part of the shoreline are taken according to a randomized and stratified sampling strategy. The shoreline is divided into four 60-cm sections as shown in Figure 2b. Normally one sample is taken from each section. The sample unit size is 10 × 10 cm. For oil analysis, three replicate samples are taken from the upper 3 cm of the sediment layer with a steel sampling box; all particles with a diameter > 10 mm are discarded from the samples. Samples for sediment-bound nutrient and microbial analysis are taken from the upper 3 cm of the sediment with a spatula. Each sample is mixed well.

Interstitial water nutrient samples are taken in sampling wells consisting of PE tubes (25 cm × 20 mm id) perforated with 2-mm holes in the lower 15 cm. The tube is installed in the sediment to a depth of 15 cm and is closed at the top with a rubber stopper (Figure 2c). The water sample is taken at falling tide when the water level is at the top of the well. Standing water in the well is removed before the water sample is taken with a pipette. The water sample is filtered (0.2 μm) immediately and frozen at −18°C until analysis.

At the end of the experiment, the sediment is removed from the basin. The inner surface of the basin, including the wave generator, is cleaned with a high-pressure washer and hot water. If necessary, detergent is used. After cleaning,

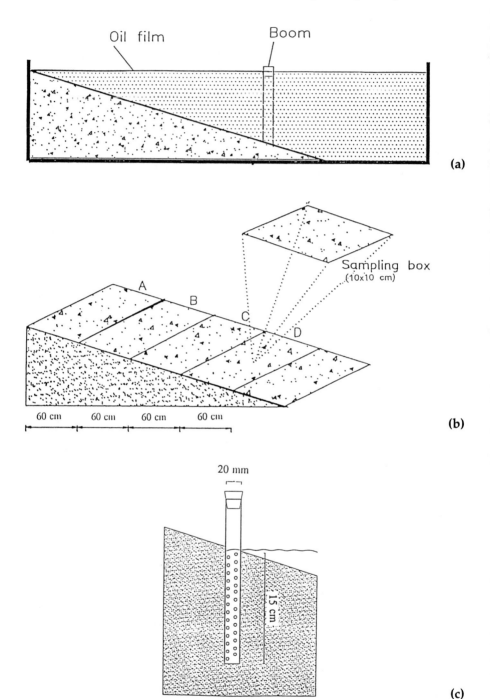

FIGURE 2. Experimental operation and sampling in the beach model experiments. (a) Oil application. (b) Sediment sampling. (c) Sampling well for interstitial water.

the basin interior is rinsed with water. New sediment is supplied to the basin for initiation of a new experiment.

MATERIALS AND METHODS

To describe the use of the basins, results from one experiment using two basins are presented: one treated with fish meal (530 g $\cdot$m^{-2}) and one without any treatment. The experiment was performed under standardized conditions (see above) with gravel beach material in the basin contaminated with topped Statfjord crude oil (150+) (2,000 mL $\cdot$m^{-2}). Additives were added 24 h after oil application.

Total heterotrophic respiration in the sediment was measured as the total amount of CO_2 produced, using a Siemens Ultramat 10 Infrared Gas Analyzer modified for septum injection. CO_2 from samples contained in sealed bottles stored in darkness at 20°C was measured after 4 h of incubation.

Concentrations of ammonium and nitrate in interstitial water and total Kjeldahl nitrogen (TKN) in sediment samples were determined by a fully automated system for nutrient analysis (Aquatec, Tecator AB), which is based on calorimetric reactions. The water samples were analyzed directly for both nitrate and ammonium, and TKN was determined after digestion with a selenium catalyzer on digester (Digestion System 40, Tecator AB) at 420°C for 120 min.

The hydrocarbons in the sediment samples (3 to 15 g) were extracted in a soxhlet extraction system (Soxtec System, Tecator) with hexane (40 mL) at 140°C for 2 h. The samples were analyzed after filtration (0.2 μm) and dilution to the appropriate concentration on gas chromatography/flame ionization detector (GC/FID) (Hewlett-Packard Series II Gas Chromatograph) with a splitless injection system.

RESULTS AND DISCUSSION

The chemical composition of the oil measured by the nC_{17}/pristane ratio changed in both basins (Figure 3). The most pronounced changes were found in the basin with fish meal, which was significantly different from the basin without treatment. It has been shown in other experiments that the nC_{17}/pristane ratio is a good indication (but not an absolute measure) for biodegradation of Statfjord crude oil as long as the ratio remains above 0.6 (Sveum and Ramstad, in preparation). Below that level, pristane is normally degraded, and thus the ratio underestimates degradation. Even within a 33-day period, addition of fish meal gives significant enhancement of oil degradation, significantly different from the control basin (student's paired t-test, Abs(Dif)-LSD: 0.05; cf. SAS 1989 for statistical details). The slight increase in the nC_{17}/pristane ratio found at the end of the experiments reflects the variability often found in sediments like this.

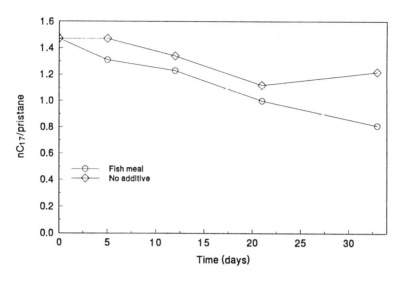

FIGURE 3. Change in the oil composition expressed with nC_{17}/pristane ratio in the basin with and without addition of fish meal.

Fish meal resulted in a high initial increase in the heterotrophic respiration, which could be observed already after one week (Figure 4). It was significantly higher than in the untreated sediment (student's paired t-test, Abs(Dif)-LSD:0.5). A decrease in the activity was found throughout the experiment, probably reflecting the degradation of the carbon added as fish meal. In the basin with no additive, the heterotrophic activity remained at a low level, with a slight increase throughout the experiment. As the level of other organic material is assumed to be very low in this sediment, the CO_2 emitted from the sediment is assumed to be mainly from the hydrocarbons. Because the heterotrophic respiration was four to six times higher with fish meal than without additives, only a fraction of the microbial activity in the treated sediment was directed toward hydrocarbon degradation; nevertheless an enhanced biodegradation of the hydrocarbons is found. This is consistent with results presented by Sveum and Ramstad (1995, this volume); which show that although the addition of external carbon (Sveum et al. 1994); results in diauxic utilization of carbon sources, the diauxy is not absolute.

The level of total soluble nitrogen in interstitial water was significantly higher in the basin treated with the fish meal (Figure 5), (student's paired t-test, Abs(Dif)-LSD:0.91). While the nitrogen level in the untreated sediment did not change throughout the experiment, a slight decrease was found in the fish meal treated sediment. The decrease in respiration followed by a decrease in soluble nitrogen in this sediment was indeed expected since the deamination of the protein is a microbial process. The ammonium concentration of the sediment was found to be linear to the total soluble nitrogen concentration basin (Figure 5). In the bioremediation cleanup operation following the *Exxon Valdez* incident, it was reported that the concentration of soluble nitrogen in the interstitial water correlated with the hydrocarbon composition (Bragg et al. 1994).

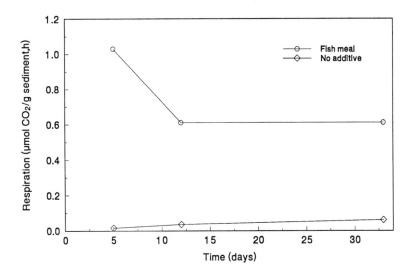

FIGURE 4. Biological activity measured as total heterotrophic respiration (emitted CO_2) in the basin with and without addition of fish meal.

These results show a significant enhancement of biodegradation of oil in a physical model of a beach sediment treated with an organic nutrient source. It has been shown that the nutrient level in the interstitial water is increased, and that sediment microbial activity increases as well. The monitored chemical and biological parameters are affected by the additive and changes in concordance

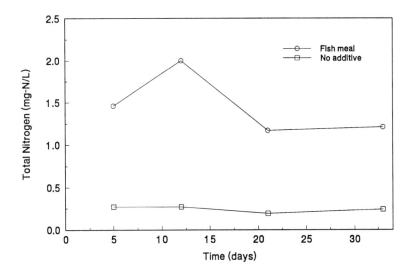

FIGURE 5. Total soluble nitrogen in interstitial water in the basins with and without addition of fish meal.

with the change in the oil composition as monitored by the nC_{17}/pristane ratio. We can thus conclude that the physical model gives biologically significant results, and that this type of open experimental system can be used to simulate biodegradation and bioremediation in natural systems.

ACKNOWLEDGMENTS

This study is part of the Esso SINTEF Coastal Oil Treatment Program (ESCOST), which is supported by Esso Norge A/S. Special thanks are due to Dr. Roger Prince, Exxon Research and Engineering, and Mr. Geir Indrebø, Esso Norge A/S, for their continuous advice and support throughout the ESCOST program.

REFERENCES

Boucher, G., and S. Chamroux. 1976. "Bacterial and meiofauna in experimental sand eco-systems. I. Materials and preliminary results." *J. Exp. Mar. Biol. Ecol.* 24: 237-249.

Bragg, J. R., R. C. Prince, E. J. Harner, and R. M. Atlas. 1994. "Effectiveness of bioremediation for the *Exxon Valdez* oil spill." *Nature* 368: 413-418.

Gibbs, G. F., and S. J. Davis. 1982. "The rate of microbial degradation of oil in a beach gravel column." *Microb. Ecol.* 3: 55.

Johnston, R. 1970. "The decomposition of crude oil residue in sand columns." *J. Mar. Biol. Assoc., U.K.* 50: 925.

McLachlan, A. 1982. "A model for estimation of water filtration and nutrient generation by exposed sandy beaches." *Marine Environmental Research* 6: 37-47.

Pugh, K. B. 1975. "A model beach system." *J. Exp. Mar. Biol. Ecol.* 18: 197.

Ramstad, S., P. Sveum, C. Bech, and L.-G. Faksness. 1995. "Modeling shoreline bioremediation; continuous flow and seawater exchange columns." In R. E. Hinchee, G. S. Douglas, and S. K. Ong (Eds.), *Monitoring and Verification of Bioremediation*, pp. 77-86. Battelle Press, Columbus, OH.

Riedl, R. J. 1971. "How much water passes through sandy beaches?" *Int. Revue Res. Hydrobiol.* 56: 923-946.

SAS. 1989. JMP® User's Guide. Version 2 of JMP.

Sveum, P., and C. Bech. 1994. "Bioremediation and physical removal of oil on shore." In R. E. Hinchee, B. C. Alleman, R. E. Hoeppel, and R. N. Miller (Eds.), *Hydrocarbon Bioremediation*, pp. 107-124. Lewis Publishers, Boca Raton, FL.

Sveum, P., L.-G. Faksness, and S. Ramstad. 1994. "Bioremediation of oil-contaminated shore-lines: The role of carbon in fertilizers." In R. E. Hinchee, B. C. Alleman, R. E. Hoeppel, and R. N. Miller (Eds.), *Hydrocarbon Bioremediation*, pp. 163-174. Lewis Publishers, Boca Raton, FL.

Sveum, P., and S. Ramstad. 1995. "Bioremediation of oil on shorelines with organic and inorganic nutrients." In R. E. Hinchee, J. A. Kittel, and H. J. Reisinger (Eds.), *Applied Bioremediation of Petroleum Hydrocarbons*, pp. 201-217. Battelle Press, Columbus, OH.

Biodegradability Study of High-Erucic-Acid-Rapeseed-Oil-Based Lubricant Additives

Enning Zhou, Alka Shanahan,
William Mammel, Jr., and Ronald L. Crawford

ABSTRACT

A variety of high-erucic-acid-rapeseed (HEAR)-oil-based lubricants, lubricant additives, and greases were examined for biodegradability at the University of Idaho Center for Hazardous Waste Remediation Research. Two standard biodegradability tests were employed, a currently accepted U.S. Environmental Protection Agency (EPA) protocol and the "Sturm Test." As is normal for tests that employ variable inocula such as sewage as a source of microorganisms, these procedures yielded variable results from one repetition to another. However, a general trend of rapid and complete biodegradability of the HEAR-oil-based materials was observed.

INTRODUCTION

Because of increasing concerns about environmental safety, there is a growing interest in developing biodegradable lubricants, especially for use in industries with high environmental impact, where soil and water contamination cannot be avoided, e.g., forestry, mining, fishing, and construction. In the event of accidental spills, the availability of biodegradable lubricants would prevent pollution of the water table, damage to aquatic life, and the destruction of indigenous microbes required for biodegradative processes. In developing environmentally friendly lubricants, we have concentrated on products based on vegetable oils and synthetic ethers, which, unlike petroleum-based products, are readily biodegradable. Here we evaluate the biodegradability of lubricant formulations containing HEAR-oil-based telomer and additives. In addition to reduced risk of environmental pollution offered by the use of HEAR oils and other polyunsaturated oils, the introduction of biodegradable lubricants and additives will also reduce waste handling and disposal problems. Increased demand for vegetable oils will also benefit agricultural producers and communities where rapeseed is a major crop.

There is no absolute standard for determining the biodegradability of lubricants. Mobil Oil Company (Cheng et al. 1991) has developed a biodegradable hydraulic oil, and has devised criteria to determine the ready biodegradability of such products: >60% conversion to CO_2 in 28 days, and aquatic toxicity > 1,000 ppm (rainbow trout). The German Environmental Agency considers a product biodegradable if conversion to CO_2 is more than 70% (Korff & Feßenbecker 1992) in 28 days. In order to be marketed with such environmentally friendly seals of approval as Germany's Blue Angel, products must satisfy the more stringent 70% Office of European Community Development requirements for the European market, and Mobil Oil's criteria for the U.S. market. The two tests used by Mobil Oil are the U.S. EPA's acute aquatic toxicity test with rainbow trout (U.S. EPA 1990), and the shake flask test (U.S. EPA 1982) for ultimate biodegradability. Since our goal is to develop a series of vegetable oil based products for use in environmentally sensitive areas that can meet these stringent criteria for biodegradability, examining the biodegradability of lubricant feedstocks is an essential step in preparing lubricant formulations with these products.

MATERIALS AND METHODS

EPA Protocol

For testing the biodegradability of lubricating oils, a protocol based on an EPA method (EPA 560/6-82/003) was used. For biodegradation of lubricants in liquid culture, 1.0 g of soil, 2.0 mL of aerated liquid inoculum, 50 mL of raw domestic influent sewage from the Moscow, Idaho, plant, and test compounds equivalent to 10 mg/L carbon were mixed in 1 L of defined liquid medium (see below), with 25 mg of casamino acids and 25 mg of yeast extract as added nutrients. This was recorded as time zero. CO_2 subsequently evolving from the biodegradation of the lubricating oils was measured periodically by $Ba(OH)_2$ titration analysis.

Tests were conducted in acclimation and biodegradation phases. The composition of the liquid medium for both phases was as follows: in 1 L of deionized water, 35 mg NH_4Cl, 15 mg KNO_3, 75 mg $K_2HPO_4 \cdot 3H_2O$, 25 mg $NaH_2PO_4 \cdot H_2O$, 10 mg KCl, 20 mg $MgSO_4$, 1 mg $FeSO_4 \cdot H_2O$, 5 mg $CaCl_2$, 0.05 mg $ZnCl_2$, 0.5 mg $MnCl_2 \cdot 4H_2O$, 0.005 mg $CuCl_2$, 0.001 mg $CoCl_2$, 0.001 mg H_3BO_3, and 0.0004 mg MoO_3.

In the acclimation phase, 1.0 g of soil inoculum, 2.0 mL of aerated liquid inoculum, 50 mL of raw domestic influent sewage, and test compounds equivalent to 4 mg/L carbon were mixed in 1 L of above defined liquid medium in a 2-L Erlenmeyer flask, with 25 mg of casamino acids and 25 mg of yeast extract as added nutrients. Test flasks were incubated in the dark at 25°C, and test compounds equivalent to 8 mg/L carbon were provided periodically to the inoculum medium before the end of acclimation phase. Only a single flask was used for each compound. During this phase, soil and sewage microorganisms were provided the opportunity to adapt to the test compounds. The medium was then used as the inoculum for the biodegradation phase.

In the biodegradation phase, 100 mL of the above inoculum was added to a specially equipped 2-L Erlenmeyer flask containing an initial concentration of test lubricating oil equivalent to 10 mg/L carbon and the same defined liquid medium, with a final volume of 1 L. Triplicate test flasks were used for each compound. A reservoir holding barium hydroxide solution was suspended in the test flask, which was sparged with CO_2-free air, sealed, and incubated with shaking (125 rpm) in the dark at 25°C. CO_2 evolved from the biodegradation of the lubricating oil was measured periodically by $Ba(OH)_2$ titration analysis. Each test also included controls receiving inoculum media but no test compounds, and controls inhibited by the addition of mercuric chloride (50 mg/L) to inhibit microbial activity.

Sturm Test

For the Sturm Test (OECD 1992), the liquid medium (pH 7.4) contained, per 1 L of deionized water, KH_2PO_4, 85.0 mg; K_2HPO_4, 217.5 mg; $Na_2HPO_4 \cdot 2H_2O$, 334.0 mg; NH_4Cl, 5.0 mg; $CaCl_2$, 27.5 mg; $MgSO_4 \cdot 7H_2O$, 22.5 mg; $FeCl_3 \cdot 6H_2O$, 0.25 mg; and suspended solids of activated sludge, 30 mg, which was used as inoculum. No casamino acid and yeast extract were added as nutrients. At time zero, test compounds (lubricants, etc.) and reference compound (canola oil, etc.) were added to test flasks to yield concentrations of 10 mg/L of carbon. Inoculum controls did not contain any additionally added test or reference compounds. The test did not have an acclimation phase. Degradation was followed for at least 28 days by monitoring the carbon dioxide produced. The CO_2 was trapped in $Ba(OH)_2$ and was measured by titration of the residual hydroxide. Biodegradations of the compounds were calculated by taking the inoculum-only control as reference. One day before the end of the test, 1 mL of concentrated HCl was added to each flask to drive off the CO_2 present in test suspensions. The last measurements were then made the following day.

HEAR-oil-based lubricants and greases were provided to the University of Idaho as "blind" samples identified only by a code. Their carbon, hydrogen, phosphorus, sulfur, and nitrogen contents were determined by standard procedures at the University of Idaho Analytical Sciences Laboratory, Holm Research Center, as shown in Table 1.

RESULTS

The product T-6000 represents a model HEAR-oil-based material. It is a lubricant additive, used as a viscosity index improver. In a biodegradability test of this lubricant (Table 2), the test compound yielded about 90% of theoretical maximum CO_2 in about 28 days by the EPA protocol (acclimation), and 60% by the Sturm Test (no acclimation). This means that a sufficient amount of the initial

TABLE 1. Lubricants, lubricant additives, and greases examined for bio-degradability.

Lubricant[a]	Elemental Composition				
	% C	% H	% N	µg/g P	µg/g S
DF-1	81.0	12.8	0.07	770	35
Waylube-68	81.7	11.7	0.08	210	10
Waylube-32	79.7	11.6	0.03	220	280
ISO VG-32	79.8	11.8	0.09	160	8
GOF239-231-T	69.3	10.5	0.12	720	29,000
05-53-1	80.2	10.9	0.13	27	93
C20000	77.1	10.2	0.03	31	32
ISO VG-68	78.9	11.7	0.04	150	12
05-66-1	76.2	11.3	0.04	21	110
FP-1	67.4	11.9	0.03	290	15
GOF-285	69.2	10.4	0.08	400	13,000
ISO VG-46	75.0	11.2	0.05	170	16
EAL-5LT	80.5	11.7	0.11	160	13
05-70-1	75.8	11.3	0.09	1,000	11,000
Waylube 220	83.2	12.2	0.04	390	420
WRL46-68	78.2	11.4	0.03	480	310
Canola oil	75.8	11.2	BDL[b]	4	5
T-6000	77.4	11.3	0.12	7	28
T-140	78.1	11.2	0.01	8	20
P-11	72.7	11.6	BDL	31,000	13
P-108	70.1	10.8	0.09	57	120,000
T-205	75.1	10.9	BDL	26	180
TP-182	77.5	11.6	0.06	6,100	49

(a) Listed lubricants, greases, and additives are all derived from rapeseed oil. Canola oil is a biodegradability control. Only some of these materials are presently used in commercial products, pending biodegradability assessments.
(b) BDL = below detection limit.

T-6000 was biodegraded to CO_2 within 28 days, which is an acceptable environmental standard, particularly by the Mobil Oil criteria. Representative results from three separate experiments are shown in Table 2 and in Figure 1.

We observed some variations in experimental results for the same lubricants between different repetitions of the measurements (e.g., compare C20000 in Figure 1A and Table 2A). We feel this was due to variations in the inocula. In

TABLE 2. Biodegradation of various lubricant materials. A, Acclimation phase; B, No acclimation phase.

A.

TEST	LUBRICANTS EXAMINED									
	Lubricant (mg C/L)									
EPA Protocol time (days)	FP-1	GOF285	C20000	T-6000	Canola oil Std	05-70-1	05-53-1	05-66-1	WRL46-68	DF-1
0.00	10.00	10.00	10.00	10.00	10.00	10.00	10.00	10.00	10.00	10.00
2.92	9.74	5.68	5.81	6.99	5.03	10.00	7.58	9.80	11.96	10.13
4.91	4.70	5.75	4.44	6.34	4.80	6.04	3.53	4.77	9.54	5.23
6.89	4.54	5.10	4.57	5.36	3.00	6.04	3.20	4.34	2.02	4.38
9.93	2.38	3.53	3.59	3.75	2.18	4.24	1.91	3.03	2.06	2.64
25.91	0.03	0.03	0.00	1.14	0.00	2.21	0.00	1.66	0.43	0.00
28.55	0.00	0.00	0.00	0.00	0.00	1.55	0.00	0.71	0.00	0.00

B.

TEST								
Lubricant (mg C/L)								
Sturm Test time (days)	T-6000	GOF285	C20000	05-70-1	05-53-1	05-66-1	WRL46-68	Canola oil Std
0.00	10.00	10.00	10.00	10.00	10.00	10.00	10.00	10.00
3.03	8.81	9.96	9.63	9.30	9.04	9.24	6.03	6.53
6.17	8.59	6.63	8.04	7.58	6.76	6.47	4.51	2.26
9.96	7.83	4.50	6.29	5.97	4.79	4.99	3.02	-0.86
13.13	7.82	4.45	6.09	5.99	4.75	4.78	2.98	-1.00
18.80	5.95	2.02	5.13	5.30	2.81	3.26	0.06	-1.00
23.97	4.37	1.13	3.91	4.23	1.94	2.23	0.00	-0.86
33.88	4.09	-0.04	3.20	3.33	1.86	1.92	0.03	-1.00
43.12	1.93	-0.04	2.41	0.98	1.07	0.94	0.00	-1.00

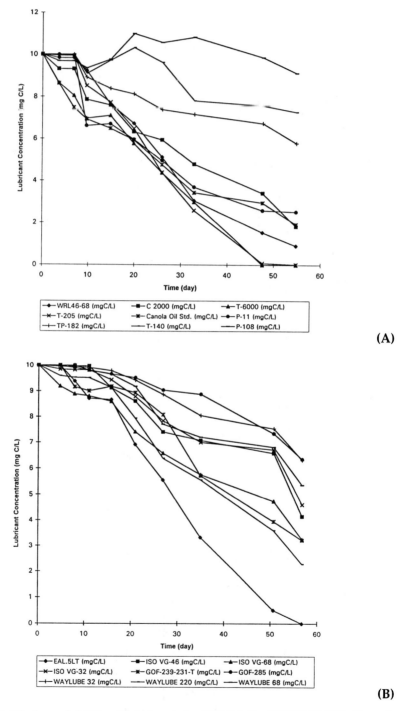

(A)

(B)

FIGURE 1. Biodegradability of lubricants by EPA Method 560/6-82/003. Experiments A and B were performed separately, with different acclimated inocula.

sewage and soils there are significant differences in composition and counts of microbial flora at different locations at different times. These differences (populations, dormancy state, characteristics of microbial communities, etc.) play a major role in the effect of microbial enzymatic activities on the lubricant substrates. An acclimation phase will not completely eliminate this source of variability.

Differences in rates of degradation of specific materials could be caused by the differences in viscosity among the products, which were pronounced, and surface tensions in the aqueous and soil environments. Less viscous oils dispersed fairly well in solution. However, the more viscous oils appeared as large oil droplets in aqueous media and might not be readily available to microorganisms.

DISCUSSION

If the Mobil Oil standard of >60% conversion of a lubricant's carbon to CO_2 in 28 days is used as an environmentally acceptable target, the following lubricants met that criterion in at least some tests: T-6000, C20000, 05-70-1, 05-53-1, 05-66-1, WRL46-68, FP-1, and DF1. None of the lubricants were as readily biodegradable as pure canola oil, but degradation rates were satisfactory for environmental protection goals.

REFERENCES

Cheng, V. M., et al. 1991. "Biodegradable and Nontoxic Hydraulic Oils." Presented at the 42nd Annual SAE Earth Moving Industry Conference and Exposition, Peoria, Illinois, April.

Korff, J., and A. Feßenbecker. 1992. "Additives for Biodegradable Lubricants." Presented at the 59th Annual meeting, National Lubricating Grease Institute, Hilton Head Island, South Carolina, October.

OECD (Office of European Community Development). 1992. *Guidelines for Testing Chemicals,* Method 301B, Modified Sturm Test, pp. 2/62-24/62.

U.S. EPA. 1982. *Chemical fate testing guidelines: aerobic aquatic biodegradation.* Method 560/6-82/003. U.S. Environmental Protection Agency.

U.S. EPA. 1990. *Methods for measuring acute toxicity of effluents and receiving water to fresh water and marine organism.* EPA 600/4-90/027. U.S. Environmental Protection Agency.

Detecting Organic Groundwater Contamination Using Electrical Resistivity and VLF Surveys

Alvin K. Benson, Kelly L. Payne, and Melissa A. Stubben

ABSTRACT

Geophysical methods can be helpful in mapping areas of contaminated soil and groundwater. Electrical resistivity (ER) and very low frequency electromagnetic (VLF EM) induction surveys were carried out at a site of shallow hydrocarbon contamination. Previously installed monitoring wells facilitated analysis of water chemistry to enhance interpretation of the geophysical data. The ER and VLF EM data correlate well and were used to map the contaminant plume, which was delineated as an area of high apparent and interval resistivities.

INTRODUCTION

Electrical resistivity (ER) and electromagnetic (EM) induction surveys are sensitive to groundwater quality and hydrocarbons in a porous medium (Telford et al. 1990; Benson 1991). Benson (1991) indicates that hydrocarbon plumes may be delineated as resistivity highs because hydrocarbons have high resistivities relative to water, or as resistivity lows if inorganic compounds are added to contaminated groundwater to stimulate bioremediation, thus increasing the total dissolved solids (TDS) in the water. Also, where the hydrocarbons are being biodegraded, resistivity values may be lower, because biodegradation tends to increase the amount of TDS in the groundwater. Studies by Foster et al. (1987) and Benson et al.(1991) report low resistivities over hydrocarbon-contaminant plumes where TDS values were increased by biodegradation and bioremediation.

Most studies have used contoured *apparent* resistivity data at a given depth to model plumes of hydrocarbon contamination, but we will use *interval* resistivities obtained by iterative computer modeling of the apparent resistivity data. Apparent resistivity is the resistivity of homogeneous, isotropic ground which would give the same voltage-current relationships as an interval of actual subsurface geology. Therefore, the measured apparent resistivity values are affected by the thickness and fluid content of each of the subsurface layers. Thus, interval resistivities should provide a more accurate picture of resistivity as a function of depth.

SITE INFORMATION

Data were collected from an alluvial basin site in central Utah Valley, Utah County, Utah. The shallow subsurface geology of the study site consists of unconsolidated to semiconsolidated alluvial deposits of gravel, sand, silt and clay (Gates & Freethey 1989). Monitoring wells (MW) previously installed at the site facilitated analyses of water chemistry and groundwater flow direction to enable correlation between geophysical data, hydrologic conditions, and water chemistry. A service station to the north of, and upgradient from, the study site reported a leak from an underground gasoline tank in 1990. Gasoline has contaminated the groundwater as far south as MW-3 and MW-4 (Figure 1) and was leaking into the stream running along the east edge of the gas station and study site. Groundwater pumped from several wells is being treated on site to remove contaminants (Keith 1993). This has arrested further migration of the contaminant plume.

Water levels in the monitoring wells were measured in May 1993, and the elevation of the water level in each well was determined relative to a chosen datum. The surface of the water table was contoured, and the gradient was calculated to be 9.9 cm per 100 m. This gradient indicates a near southwesterly flow direction toward Utah Lake.

WATER CHEMISTRY

Water samples were collected from the monitoring wells in October 1993. Samples from each well were analyzed at Brigham Young University (BYU) for anion concentrations using an ion chromatograph and for cation concentrations using an atomic absorption spectrometer. Using a gas chromatograph, ground-waters from MW-1 and MW-3 were analyzed for total petroleum hydrocarbons (TPH) and benzene, toluene, ethylbenzene, and xylenes (BTEX) by Richard Laboratories in Pleasant Grove, Utah.

Results of the water analyses are summarized in Tables 1 and 2. MW-3 and MW-4 encounter the contaminant plume, while MW-1 had no detectable TPH or BTEX, and MW-2 had no detectable product odor. From the direction of groundwater flow, contamination might be expected in all of the monitoring wells. However, the fact that contamination is not present in MW-1 and MW-2 indicates that pumping began soon enough at the site to prevent the contaminant from migrating extensively, and/or perhaps the shallow subsurface layers have a low hydraulic conductivity that has slowed the movement of the contaminant through the soils.

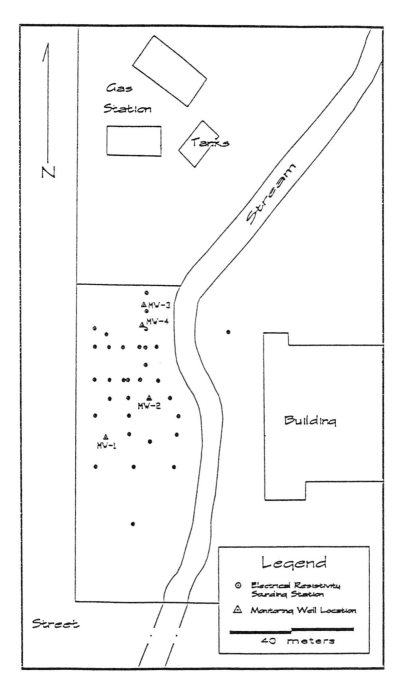

FIGURE 1. Case study site location in Provo, Utah. Map shows position of monitoring wells (MW), resistivity sounding locations, and buried gasoline tanks.

TABLE 1. TDS analysis results for groundwater samples collected from study site (in mg/L).

Monitoring Well	F⁻	Cl⁻	NO₃⁻	SO₄²⁻	HCO₃⁻	Na⁺	K⁺	Ca²⁺	Mg²⁺	TDS
MW-1	0.1	49	1.9	69	375	34	4.4	116	25	675
MW-2	0.3	40	1.0	68	384	26	3.8	119	26	668
MW-3	0.4	44	0.4	45	743	36	16.5	119	46	1,121
MW-4	0.7	46	0.5	54	889	49	16.0	199	53	1,307

ELECTRICAL RESISTIVITY RESULTS

Between 6 October and 13 October 1993, 32 electrical resistivity soundings were made (Figure 1). To obtain maximum coverage, electrode arrays were set up north-south and east-west, and for some soundings the array crossed the stream. Some of the soundings were repeated on 3 December 1993 to ensure reproducibility and accuracy of the data.

The apparent resistivity data collected in this survey were modeled to obtain interval resistivities, which were plotted and contoured for different electrode spacings (a-spacings). The contour map for the 4-meter a-spacing (Figure 2) represents the interval resistivity values at a depth near the water table. Resistivities are high in the northern part of the study area where contaminant levels are high. These values decrease to the south where contaminant levels drop below detectable limits. Resistivity values range from 36 ohm-m near MW-3 to less than 11 ohm-m near MW-1. The contaminant plume appears to be outlined by a geometrical-shaped "nose" of high resistivity values corresponding to the area of high contamination as determined from the water chemistry analyses.

The areas of high resistivity correspond to a higher TDS concentration (Table 1), which should decrease the resistivity of the water table. At this site, however, the resistivity appears to be controlled primarily by gasoline floating

TABLE 2. BTEX and TPH analysis results for groundwater samples collected from study site (in mg/L).

Monitoring Well	Benzene	Toluene	Ethylbenzene	Xylenes	TPH
MW-1	<5.0	<5.0	<5.0	<5.0	<500
MW-3	2,850	344	414	4,432	13,258

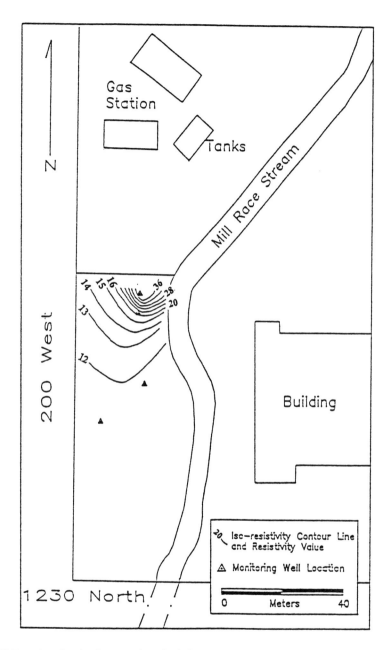

FIGURE 2. Study site interval resistivity contour map at a depth of 1.6 meters. Contour interval is 4 ohm-m to the north and 1 ohm-m to the south.

on the water table and/or gasoline constituents wetting the soils above the water table, rather than by TDS in the contaminated water. This effect may be further enhanced by clays at the site which can trap the gasoline and prevent its escape.

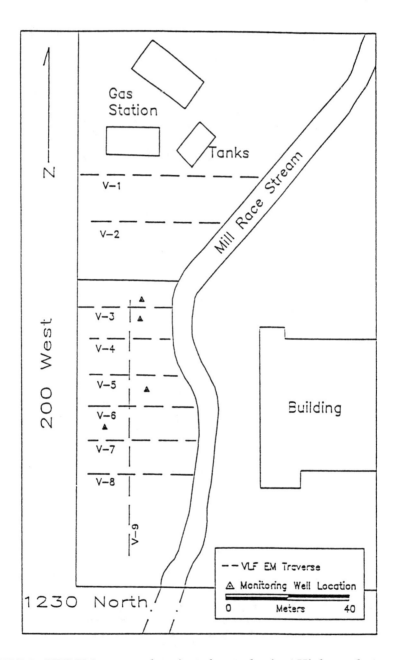

FIGURE 3.   VLF EM traverse locations for study site.   Higher values on the
vertical axis correspond to lower resistivities.

At other sites in Utah and Arizona, where the TDS increases relative to the amount
of residual hydrocarbons (due to degradation of hydrocarbons), groundwater contamination has been mapped as a resistivity low (Benson et al. 1991; Benson 1992).

VLF SURVEY RESULTS

VLF data were collected along nine traverses on 29 November and 3 December 1993 (Figure 3). These data were filtered, processed, and plotted at a selected depth near the water table (Figure 4).

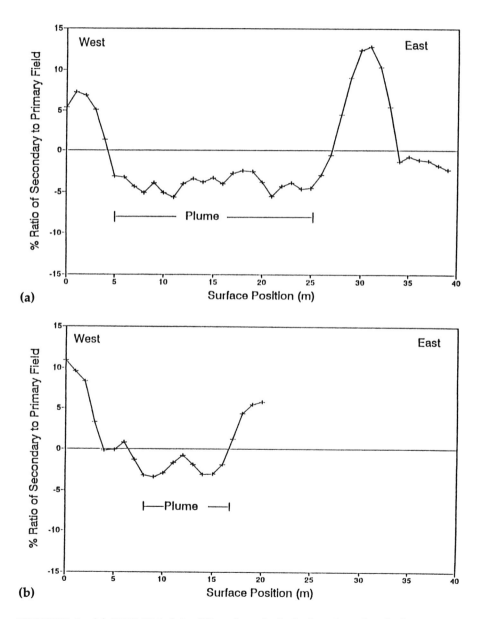

(a)

(b)

FIGURE 4. (a) VLF EM data filtered and plotted at 1-m depth for traverse V-1. (b) VLF EM data filtered and plotted at 1-m depth for traverse V-3.

Interpretation of these traverses indicate a resistivity high (less positive values) over the contaminant plume, correlating well with the ER data. The magnitude of the resistivity of the subsurface materials decreases going south from the gas station, or away from the contaminated area. VLF traverses 1 and 3 (Figure 4a and 4b) cross the contaminant plume. These figures show the contaminant plume as a resistivity high surrounded on both sides by materials of lower resistivity. In Figure 4a, the resistivity high begins about 5 m from the west edge of the study area and extends to about 26 m from the west edge. Figure 4b shows a resistivity high beginning about 7 m from the west edge of the study area and extending to abut 16 m from the west edge. Traverses 4 through 9 also show decreasing contamination toward the south.

CONCLUSIONS

ER and VLF surveys were successfully combined to help outline a contaminant plume produced by the leaking gasoline. Good correlation exists between the geophysical surveys and hydrocarbon contamination in strategically located wells, and the geophysical data have effectively extended the horizontal extent of the borehole data. Both geophysical methods identified the contaminant plume by high resistivity values, even though there were higher TDS concentrations in the contaminated groundwater as compared to the cleaner groundwater. The resistivity of the hydrocarbons in the soils and those floating on the water table appears to be the dominating factor at this site. Interpretations of the filtered VLF data correlate well with the interpreted interval ER data. These geophysical methods show promise for application at sites of hydrocarbon contamination to help identify the source(s) of contamination, strategically locate monitoring wells, and monitor cleanup activities.

REFERENCES

Benson, A. K. 1992. "Integrating seismic, resistivity, and ground penetrating radar to delineate the water table and ground-water contamination." In: Y. K. Kharaka and A. S. Maest (Eds.), *Water-Rock Interaction, V.1, Low Temperature Environments*. Balkema, Rotterdam, Netherlands. pp. 361-365.

Benson, A. K., C. Frederickson, and N. B. Mustoe. 1991. "Ground-penetrating radar, electrical resistivity, soil and water quality studies integrated to determine the source(s) and geometry of hydrocarbon contamination at a site in north-central Arizona." In: J. P. McClapin (Ed.), *Proceedings of the 27th Symposium on Engineering Geology and Geotechnical Engineering*. Utah State University, Logan, UT. pp. 38.1-38-13.

Benson, R. C. 1991. "Remote sensing and geophysical methods for evaluation of subsurface conditions." In: D. M. Nielsen (Ed.), *Practical Handbook of Ground-Water Monitoring*. Lewis Publishers, Chelsea, MI. pp. 143-194.

Foster, A. R., M. D. Veatch, and S. L. Baird. 1987. "Hazardous Waste Geophysics." *Geophysics, The Leading Edge 6*: 8-13.

Gates, J. W., and G. W. Freethey. 1989. "The relation of the geohydrologic setting to the potential for ground-water contamination in Utah." In: G. E. Cordy (Ed.), *Geology and*

Hydrology of Hazardous Waste, Mining-waste, and Repository Sites in Utah. Utah Geological Association, Salt Lake City, UT. pp. 11-28.

Keith, J. D. 1993. Personal communication. Brigham Young University, Provo, UT.

Telford, W. M., L. P. Geldart, and R. A. Sheriff. 1990. *Applied Geophysics,* 2nd ed. Cambridge University Press, New York, NY.

Continuous Bioventing Monitoring Using a New Sensor Technology

Dong X. Li

ABSTRACT

Vadose zone oxygen sensors can be used effectively to improve bioventing operation and monitoring. The capabilities of the oxygen sensors for continuous oxygen measurements unattended can improve the current methods and offer alternative approaches for respiration measurements. These sensors have been used to develop a new dynamic technique of evaluating in situ respiration rates during air injection or vapor extraction which has several advantages over the traditional static oxygen uptake method. By using a subsurface oxygen sensor, the dynamic technique offers continuous monitoring capability during the bioventing process. Unlike the traditional respiration test that measures localized respiration rates, this method determines an average respiration rate in the air flow path. Because the measurements can be made at the startup of a remediation process, the operation can run without interruption. The application of this new technique and its advantages are documented in three sites.

INTRODUCTION

Monitoring is an important part of environmental remediation, especially for in situ remediation processes such as vapor extraction or bioventing. Gaseous components in a soil system are the most mobile and thus easiest to monitor. In bioventing processes, in situ biodegradation of hydrocarbon contaminants in soil is most easily enhanced by providing oxygen to the subsurface soil environment through either air injection or extraction. The biological activity in soil can be most easily monitored by measuring changes in oxygen and carbon dioxide contents in the soil gas (Thornton & Wootan 1982). It has been shown that soil gas oxygen is a more reliable indicator than carbon dioxide because of possible inorganic sources and sinks of carbon dioxide in the soil (Dupont et al. 1991; Hinchee & Ong 1992).

Currently most soil gas monitoring involves sampling and analysis at one discrete moment and point in time. This type of monitoring is not only labor intensive and expensive, but also prone to error. A subsurface oxygen sensor recently introduced (by Datawrite Research Co.). The oxygen sensor consists of an electrochemical cell, which is capable of performing continuous oxygen measurements with minimum disturbance to the subsurface condition. The laboratory and field testing of the oxygen sensor showed that it is ideally suited for bioventing monitoring (Li & Lundegard 1995). The application of this subsurface oxygen sensor in bioventing monitoring is described and illustrated using field data from three different sites.

SITE DESCRIPTION

Site A is a former bulk petroleum facility. The subsurface sediments consist predominately of silt, with the depth to groundwater at 60 ft (18 m). A well-defined diesel hydrocarbon plume was delineated between 15 and 25 ft (5 and 8 m) below the surface. Three oxygen sensors were installed at three different depths (16 ft, 23 ft, and 33 ft) near the center of the hydrocarbon plume. The sensors were packed with sand and isolated from each other and the surface with grout (Figure 1a). A PVC pipe was installed in the same borehole with the oxygen sensors, which is screened between 20 and 25 ft (6 and 8 m) below the surface within the hydrocarbon contaminated layer. Three additional oxygen sensors were installed in a similar fashion also in three different depths (17.5 ft, 25 ft, and 33 ft) in a nearby location outside the hydrocarbon plume with a PVC pipe screened at the bottom between 30 and 35 ft (9 and 11 m). Oxygen concentration readings from all six sensors immediately following the installation are shown in Figure 2.

Site B is a former service station. The soil consists of fine-grained to medium-grained sand and silty sand with a groundwater level at about 200 ft (60 m) below grade. A gasoline hydrocarbon plume was identified between 30 and 60 ft (9 and 18 m) below the surface. Oxygen sensors were hung in the three existing wells near the screen interval (30 to 65 ft or 9 to 20 m) (Figure 1b). The advantage of this method of sensor installation is that the sensors can be easily and rapidly installed and then retrieved and reused in other locations. However, a long gravel packed screened interval found in many wells can act as an easy conduit for subsurface gas movement, causing to considerable concentration fluctuation (Figures 5 and 7), where a significant diurnal variation is present. Therefore, for a more accurate soil gas measurement in many cases, a direct installation is preferred (Figure 1a). Five other oxygen sensors were directly installed at depths ranging from 20 to 40 ft (6 to 12 m) below the surface at Site B.

Site C consists of several operating petroleum hydrocarbon pipelines. The soil sediments consist mostly of sand with the groundwater depth at approximately 10 ft (3 m). Hydrocarbon contamination found in the soil includes degraded gasoline and diesel. An active vapor extraction system has been operating at the site for more than 2 years through more than 30 vapor extraction wells. The hydrocarbon concentration level has dropped below 10 ppm in extracted gas stream for every well. All the wells are screened from 5 to 15 ft (1.5 to 4.5 m). Oxygen sensors were installed in all the wells (Figure 1b).

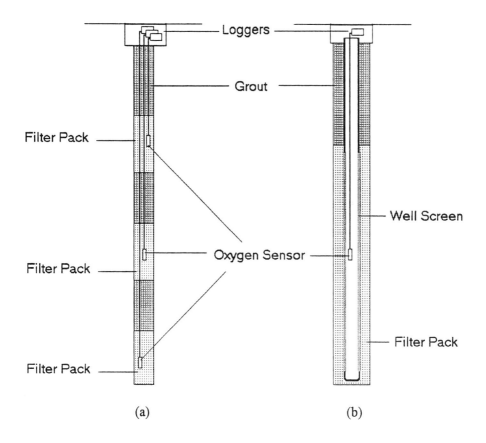

FIGURE 1. Schematic of two sensor installation variations: (a) Direct installation in soil; (b) Installation in a normal vapor extraction/monitoring well. Method shown in (a) is recommended for accurate measurement of zone-specific data.

RESPIRATION MEASUREMENT

Measurements of oxygen depletion and carbon dioxide enrichment in soil regions have been used to locate subsurface hydrocarbons (Ririe & Sweeney 1993). This same approach can be used to define an oxygen-limited soil region amenable for bioventing either through a soil gas survey or subsurface oxygen sensor measurements. The background oxygen concentrations are approximately 10% at Site A, and within the plume are near 0%, indicating that an oxygen-limited condition exists at Site A (Figure 2).

A higher oxygen concentration, which varied between 1% and 2%, was found below the hydrocarbon plume indicating that oxygen from ground surface can still reach the bottom of the plume through natural dispersion. The oxygen concentration variation correlates well with the variation of differential pressure between

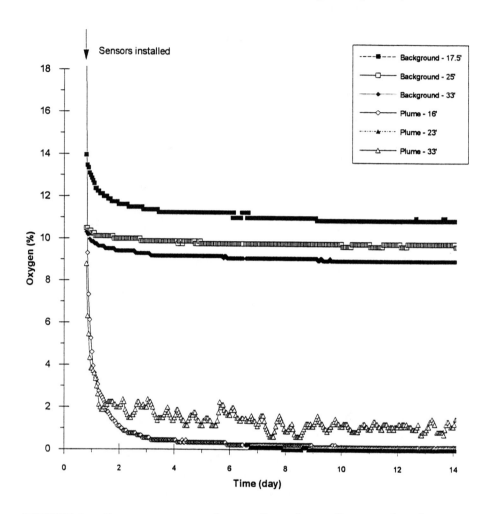

FIGURE 2. Oxygen concentration readings from all sensor locations at a former bulk petroleum facility (Site A) after installation. Note the low oxygen readings inside the hydrocarbon plume and a small diurnal variation for the sensor below the hydrocarbon plume.

ambient and subsurface soil gas (Li 1994), which relates directly to surface atmospheric pressure fluctuations (Falta et al. 1994). However, the magnitude of soil gas movement driven by such pressure variation is very limited (inches or less) for a typical soil permeability value. The magnitude of oxygen concentration variation is an indication of the concentration gradient near the sensor location.

Petroleum hydrocarbons are known to be readily biodegradable in most soils, especially under aerobic conditions (Davis 1967). The rate of aerobic biodegradation can be determined through an in situ respiration measurement (Hinchee & Ong 1992). In situ respiration normally is measured from the rate of oxygen depletion in a static soil condition. If the initial oxygen concentration in the soil is uniform, a linear concentration decline is expected for a zero-order rate. However, because oxygen most often is severely depleted in a hydrocarbon-contaminated soil, oxygen level in the soil has to be first elevated through some form of air injection or vapor extraction before a respirometry test can be performed.

Diffusion Effect on Respiration Measurement

A soil region with an elevated oxygen level created by air injection has finite size. The oxygen concentration will decline over time by natural gas diffusion even without respiration, as illustrated by air injection at the background location at Site A (Figure 3). After 10 hours of air injection at 100 ft^3/hr (cfh) (2.8 m^3/h), only the oxygen sensor near the screened interval of the air injection pipe showed significant increase in oxygen concentration, indicating a limited region of elevated oxygen concentration. Depending on the size of oxygen plume, a window could be available to perform a static respiration measurement with minimum interference from the diffusion effect (Figure 3).

The 10-hour air injection test was repeated at the plume location at Site A (Figure 4). During the sensor installation, it was noted that the soil in the diesel-contaminated layer was much tighter than the uncontaminated soil. Because the injection pipe is screened within the plume layer, a significant pressure build-up was observed that reached 5 psig at the beginning of the air injection and gradually settled at 2 psig at the end of the injection. Because the oxygen sensor measures the partial pressure of oxygen, apparent higher-than-ambient-level oxygen readings from the sensor next to the screened section of the injection pipe is a result of the pressure buildup during the air injection. A respiration rate of approximately 3%/day was measured from the initial time-concentration slope within first 10-h window of the pump shutdown. The slope after the initial period as well as the slopes from two other sensors are much higher (Figure 4) primarily due to the diffusion effect, which can lead to overestimation of the respiration rate.

The length of the window, or the lag period, is a function of gas diffusivity and gas plume size. Only the gas plume size can be modified easily by selecting the injection flowrate and duration. The effect of plume size on respiration rate measurement is illustrated by two in situ respirometry tests using the same air injection well at Site B (Figure 5). Rates of oxygen uptake in the injection well and a subsurface oxygen sensor 15 ft from the injection well were measured after 3 weeks of air injection in the first test and 1 week of air injection in the second test. The air injection rate was 200 cfh (5.7 m^3/h) for both tests. Both tests showed a very similar initial rate of oxygen uptake (4%/day) measured at both the injection well and the oxygen sensor. However, the rate of oxygen uptake deviates gradually from the expected linear respiration rate. A larger deviation was observed for the shorter air injection period (Figure 5). This non-linear effect is believed to be caused by natural soil gas diffusion. In this case at Site B, the oxygen uptake measurement will likely underestimate the actual respiration rate as a result of the diffusion effect.

Significant oxygen concentration variations were observed only in the injection well, but not at the subsurface soil sensor location. The concentration variations

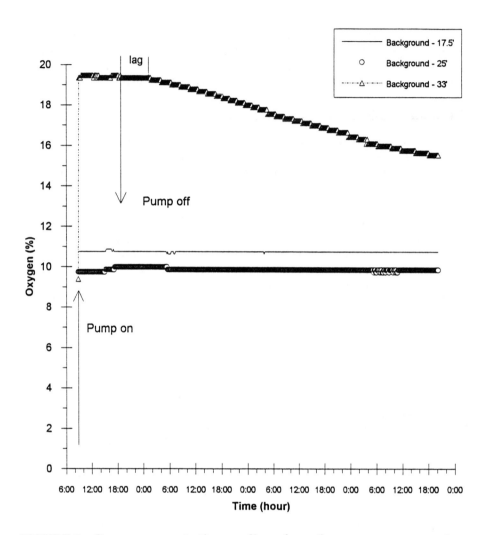

FIGURE 3. Oxygen concentration readings from three oxygen sensors in a background location during a respirometry test at a former bulk petroleum facility (Site A). Note a lag period prior to a gradual decline of oxygen concentration as a result of diffusion.

in the well were also driven by the diurnal pressure variations (Li 1994). Vertical soil gas movement in the well through the long screened interval is believed to cause the concentration fluctuations. Therefore, the direct-installed subsurface sensor provides more reliable oxygen measurements.

An Alternative Method for Respiration Measurement

When air is injected into a soil region with active aerobic biodegradation, oxygen is being consumed even during the air injection period. As the duration

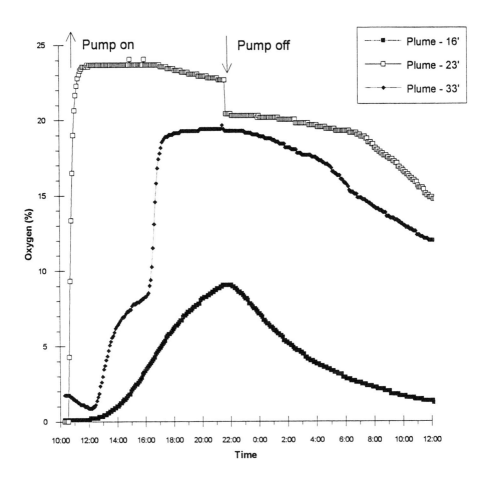

FIGURE 4. Oxygen concentration readings from three oxygen sensors in a hydrocarbon plume during a respirometry test at a former bulk petroleum facility (Site A). Note the air injection pressure causing higher-than-ambient-level oxygen readings during air injection and a sharp drop in oxygen readings when the pump was turned off.

of air injection increases, the oxygen concentration profile in the soil approaches a steady state. This steady-state condition is a function of average respiration rate, R, and the average time, t, for air to travel from injection point to a sensor location:

$$C_{ss} = C_{in} - R\,t \qquad (1)$$

where C_{in} and C_{ss} are oxygen concentrations of injected air and steady state at a sensor location, respectively. The average air travel time, t, is inversely proportional to the injection flowrate and is a proportional function of the distance between injection point and the sensor location. The oxygen concentration profile

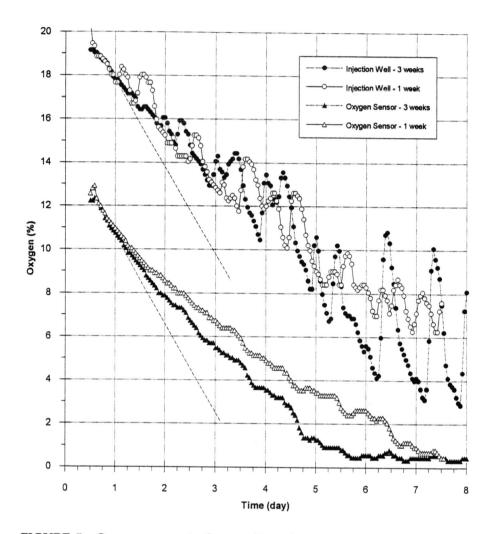

FIGURE 5. Oxygen concentration readings from an oxygen sensor in the injection well and an in situ sensor in a hydrocarbon plume during a respirometry test at a former service station site (Site B). Note the much larger diurnal concentration fluctuations found in the well compared to that in the soil measured by the in situ oxygen sensor.

can vary depending on the airflow pattern. To acquire meaningful data from post-injection respiration measurement, air injection rate has to be sufficiently high to minimize the spatial oxygen concentration gradient or the diffusional effect. Alternatively, the respiration rate is measured dynamically by rearranging Equation 1 to:

$$R = (C_{in} - C_{ss})/t \qquad (2)$$

where t is the average air travel time or the time required to fill the pore volume between injection point to sensing point at a given air injection rate. This time value can be evaluated from a time-concentration breakthrough curve measured by subsurface oxygen sensors within the radius of influence at the initial stage of air injection. A quick evaluation of the average time can be approximate as the time when oxygen concentration reaches the average of the initial and steady-state concentrations, $(C_0 + C_{ss})/2$ (Li 1995a).

As an example, the average air travel time from the injection well to an oxygen sensor 15 feet from the well roughly 33 hours at Site B, based on 200 cfh (5.7 m^3/h) air injection rate, 20.9% injection oxygen concentration (C_{in}), 0.3% initial concentration (C_0), and 14% steady-state concentration (C_{ss}) (Figure 6). An average respiration rate of about 5%/day is calculated from Equation 2, which is in near agreement with the 4%/day local respiration rates measured by the

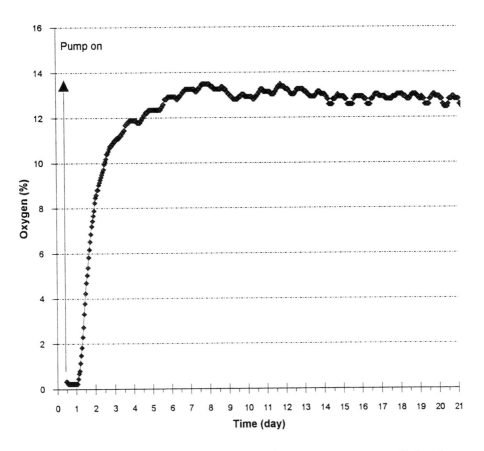

FIGURE 6. Oxygen concentration readings from a oxygen sensor 15 feet from the air injection well during a bioventing in situ respirometry test, at a former service station site (Site B).

subsurface sensor and the sensor in the injection well based on the static of oxygen uptake (Figure 5).

This method imposes little restriction on the air injection rate as long as sensing points are within the radius of influence. At a constant air injection rate, the injection oxygen concentration and the average air travel time in the soil normally do not change significantly over time. According to Equation 2, the respiration rate can be continuously monitored by measuring oxygen sensor readings without any interruption to the bioventing process. We have found through continuous biodegradation rate measurements in our biofiltration studies that the biodegradation rate of hydrocarbons seldom stays unchanged with changing conditions. The dynamic respiration measurement using subsurface oxygen sensors provides a continuous monitoring technique for biodegradation rate which will improve the current method of bioventing system optimization and operation.

Respiration Measurement During Vapor Extraction

Bioventing has long been recognized as a side benefit of vapor extraction (Thornton and Wootan 1982). Both static and dynamic respiration measurement can be performed in a vapor extraction setup. By continuously monitoring changes of oxygen concentration in the extracted stream at the startup of a vapor extraction process, an average respiration rate within the captured zone can be calculated. Representative oxygen concentration measurements from a monitoring well (DMW-7) and two of the vapor extraction wells (SVE-9 and SVE-23) at Site C are shown in Figure 7.

Large oxygen concentration fluctuations were observed in DMW-7 with a fluctuation cycle of roughly 24 hours. The large magnitude of concentration variations (as much as 8% in one day) is believed to be a result of gas movement in the well driven by the diurnal pressure variations. These variations make accurate respiration measurement difficult. Changes in oxygen concentration in response to changes in vapor extraction operation suggest that DMW-7 is within the radius of influence of the vapor extraction system. A sharp (5%) drop in oxygen concentration in DMW-7 when the vapor extraction pump was turned off and a rapid concentration rise when the pump was turned back on indicate that a high respiration rate (at least 5%/day) exists around DMW-7.

Moderate diurnal variations (about 2% in one day) were found in SVE-9, only when the vapor extraction pump was turned off (Figure 7), because the vacuum pressure is much greater than the atmospheric pressure variation. A jump in oxygen concentration when the pump was turned off is a result of the removal of the vacuum pressure or an apparent increase in oxygen partial pressure, which does not affect the oxygen uptake measurement. The pressure effect on sensor readings is proportional to the overall pressure, which can be easily adjusted for the pressure change. A respiration rate of approximately 1.5%/day was measured from an averaged static rate of oxygen uptake. An average respiration rate of 4%/day for the captured zone of SVE-9 was calculated using Equation 2.

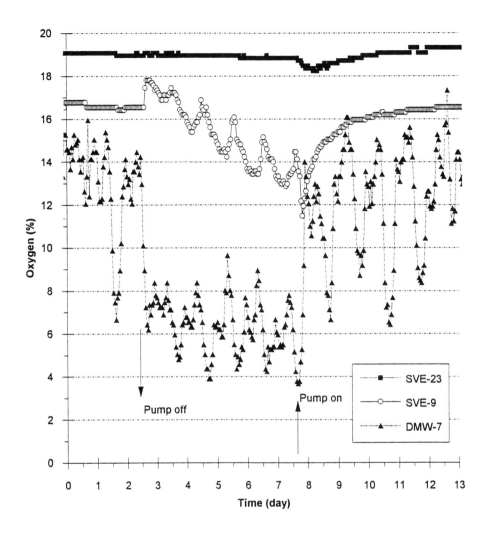

FIGURE 7. Oxygen concentration readings from oxygen sensors in two vapor extraction wells (SVE-9 and SVE-23) and a monitoring well (DMW-7) during an in situ respirometry test at a pipeline corridor site (Site C).

No detectable diurnal variation was found in SVE-23. A very low respiration rate of roughly 0.05%/day was determined near SVE-23 from the oxygen uptake measurement. An initial oxygen concentration drop in SVE-23, when the vapor extraction pump was turned back on, suggests that a soil region away from SVE-23 in the captured zone has a higher rate of respiration (Figure 7). An average respiration rate of 0.25%/day was calculated using Equation 2.

Local respiration rates from both vapor extraction wells (SVE-9 and SVE-23) were lower than the average respiration rates in the captured zone. The lower respiration rate near vapor extraction wells is not totally unexpected, because many of the volatile hydrocarbons have been removed in 2 years of vapor extraction operation (especially in the soil region near the vapor extraction wells), which is consistent with the high respiration rate found near the monitoring well (DMW-7). Because of this nonhomogeneous distribution of biologically active regions created by the vapor extraction process, the average respiration rate is a more representative measurement, and because the measurements can be made during startup, it allows the process to run without interruption.

REFERENCES

Davis, J. B. 1967. *Petroleum Microbiology*. Elsevier Publishing Co., New York, NY.

Dupont, R. R., W. J. Doucette, and R. E. Hinchee. 1991. "Assessment of In Situ Bioremediation Potential and the Application of Bioventing at a Fuel-Contaminated Site." In: R. E. Hinchee and R. F. Olfenbuttel (Eds.), *In Situ and On-Site Bioreclamation*. Butterworth-Heinemann, Stoneham, MA. pp. 262-282.

Falta, R. W., J. Rossabi, B. D. Riha, J. D. Schramm, and B. B. Looney. 1994. "Quantitative Atmospheric/Subsurface Pressure Relationships: Theory and Practice." Presented at the AGU 1994 Fall Meeting, San Francisco, CA, December 5-9, 1994, H41B-6.

Hinchee, R. E., and S. K. Ong. 1992. "A Rapid In Situ Respiration Test for Measuring Aerobic Biodegradation Rates of Hydrocarbons in Soil." *J. Air Waste Manage. Assoc., 42*: 1305-1312.

Li, D. X. 1994. "Monitoring Natural Soil Venting Processes Using Innovative Sensor Technology." Presented at the AGU 1994 Fall Meeting, San Francisco, CA, December 5-9, 1994, H41B-4.

Li, D. X. 1995. "Bioventing Feasibility Assessment and System Design Using Subsurface Oxygen Sensors." In review.

Li, D. X. 1995a. "Interpretation of In Situ Sensor Data Using Transport Models for Porous Media." Unpublished work.

Li, D. X., and P. D. Lundegard. 1995. "Evaluation of Subsurface Oxygen Sensors for Remediation Monitoring." In review.

Ririe, G. T., and R. E. Sweeney. 1993. "Comparison of hydrocarbon gases in soils from natural seeps and anthropogenic sources." *Proceedings of the 1993 Petroleum Hydrocarbons and Organic Chemicals in Ground Water: Prevention, Detection, and Restoration*, Houston, TX. pp. 593-607.

Thornton, J. S., and W. L. Wootan, Jr. 1982. "Venting for the removal of hydrocarbon vapors from gasoline-contaminated soil." *J. Environ. Sci. Health, A17*(1): 31-44.

Degradation Tests with PAH-Metabolizing Soil Bacteria for In Situ Bioremediation

Georg Maue and Wolfgang Dott

ABSTRACT

A rapid screening test for PAH degradation was used to evaluate the metabolizing potential of a bacterial community from a contaminated soil. The test was performed on a small scale within a few days using direct fluorometric quantitative analysis of selected PAHs. Therefore, a wide range of isolates and mixed cultures could be investigated under various substrate conditions with little time and material expenditure. Furthermore, the composition of the bacterial community after growth on different carbon sources was observed. The tests accompanied PAH degradation experiments in a bioreactor for the detection of suitable soil bacteria for in situ bioremediation. A mixed culture consisting of at least five different bacterial species was found in samples of the bioreactor. Different precultivation substrates (PAH) did not influence the stability of the bacterial community. Although only a few isolates metabolized single PAHs (acenaphthene, anthracene, phenanthrene) as sole substrates, the mixed culture metabolized these PAHs within a few days regardless of the precultivation. The stability of the mixed culture indicates its resistance to substrate changes that may occur during in situ bioremediation processes. Enhanced degradation rates occurred following the growth on acenaphthene and phenanthrene.

INTRODUCTION

In recent years bacterial mixed cultures and selected strains with a high capacity for PAH degradation commonly have been obtained by cultivating these microorganisms from contaminated soils. Bacteria can develop particular metabolizing potentials specially after long periods of adaptation (Kästner et al. 1991; Cerniglia 1992), and those bacteria can be used for biological decontamination. However, the degradation potential often gets lost when environmental conditions change. The available substrates influence plasmid expression in bacteria, as well as the composition of the mixed culture due to different species tolerance.

Hence, the presence of a stable bacterial community with a high degradation potential is desirable for its practical use in bioremediation processes. Therefore, a laboratory degradation test should consider different substrate conditions.

In previous investigations, a great number of species with the potential for PAH degradation could be found in reactors with high degradation rates for PAH and other hydrocarbons. This number decreased following changes in the substrate or dilution rate. The population diversity reduced probably due to the fact that only a few organisms could tolerate varied environmental conditions. On the other hand, a stable composition of the microflora indicated constant degradation rates (Maue et al. 1994b). A rapid test system with quantitative analysis (Maue et al. 1994a) allowed the quick assessment of all isolates and mixed cultures. In contrast to other (quantitative) screening tests (Kiyohara et al. 1982, Heitkamp et al. 1988), which use PAH-layers on agar plates, changes in degradation rates of selected isolates can be detected in this system. During several degradation experiments with PAHs and other hazardous compounds, BIOPRACT GmbH (Berlin, Germany) cultivated mixed cultures from different contaminated sites. Intensive degradation of 16 PAHs (16 PAHs regulated by the U.S. Environmental Protection Agency) was performed in bioreactors (Dreyer et al. 1995). Our work investigated this mixed culture from a coal tar-contaminated site. The impact of different substrate conditions on the bacterial mixed culture capable of increased PAH degradation was tested. The observed PAH degradation, as well as the changes in the composition of the bacterial community after growth on different substrates, served as criteria for monitoring the stability of the bacterial population.

EXPERIMENTAL PROCEDURES AND MATERIALS

Bacterial Cultures

The bacterial mixture was obtained from a tar oil-contaminated site (including a wide range of PAH) and has proven its tar oil converting capacity, and PAH converting capacity, in degradation experiments carried out in the laboratory of the BIOPRACT GmbH (Dreyer et al. 1995). The mixed culture was kept frozen for long-term storage. For use as inoculum it was cultivated in a 300-mL shaking flask with tar oil as substrate for 3 days, and subsequently transferred into the degradation bioreactor (10% v/v). During these degradation experiments, samples were taken from the inoculum (for additional investigations with substrate changes), and from batch-start (after a 16-h fermentation process) and batch-end (after a 7-d fermentation process) for isolation and identification of the bacteria and for degradation tests with the isolates. For identification tests (Kämpfer et al. 1991), isolates from the liquid culture were cultivated on R2A (Difco) agar plates. The identification is based on 87 physiological tests (biochemical tests, carbon source utilization, sugar fermentation and qualitative enzyme tests), which are performed in microtitration plates (Greiner, Nürtingen, Germany). Additionally, hydrocarbon-agar and diesel-agar (Steiof 1993) were used for cell counting.

Cultivation on Different Carbon Sources

Five mL of the mixed culture inoculum (OD_{600} = 0.1) was added to 95 mL mineral salt medium (MSM) (Maue et al. 1994a) in sterile Erlenmeyer flasks in addition to the autoclaved carbon source (10 mg acenaphthene, phenanthrene, pyrene, benz(k)fluoranthene, or 100 mg glucose in sterile solution; these concentrations were chosen to allow maximum growth). Substrate-free medium (5 mL of mixed culture inoculum in 95 mL MSM) served as a control. After 7 days of incubation (110 rpm, 25°C), cell counts and degradation tests were performed. Agar plates with 30 to 100 colonies were chosen for the determination of colony distribution.

PAH Transformation Screening

Acenaphthene (2 mg/L), phenanthrene (2 mg/L) and anthracene (0.1 mg/L) were used as test substrates. PAH consumption in test tubes (5 mL) with bacteria (isolates from R2A-medium in MSM solution with an OD600 of 0.1; artificially made mixed cultures M1, M2, and M3: mixtures of isolates corresponding to their origin [inoculum, batch-start, batch-end], similarly dissolved in MSM; mixed cultures in liquid sample: 0.5 mL of precultivation samples in 4.5 mL MSM) was measured daily over 7 days (60 rpm, in the dark, 25°C). Fluorometric PAH detection (of cyclohexane extracted samples) was performed using a synchronous scan in specific wavelength distances (Maue et al. 1994a), which saves a great deal of work (no separation required).

RESULTS

Twelve isolates out of three samples from the batch fermentation process were investigated. Four different species could be identified. Seven isolates were identified as *Pseudomonas aeruginosa* (isolates 1,2,6,7,8,9,12), one as *Pseudomonas putida* biotype A (4), one as *Alcaligenes xylosoxidans ssp. denitrificans* (5), one as *Comamonas testosteroni* (10), and two could not be identified (3,11).

Bacterial Growth on Different Carbon Sources

The tested mixed culture showed fast growth on tar oil compounds in the batch reactor after inoculation. The number of colony-forming units (CFUs) increased by two orders of magnitude within 16 h (see Table 1). The total cell counts on R2A media were similar to those on diesel or hydrocarbon agar (except for one count on hydrocarbon agar, Table 1).

The total number and diversity of the colonies were not influenced by most of the precultivation procedures of the inoculum sample on different PAHs (see Table 2). Only acenaphthene and glucose stimulated the bacterial growth (1 to 2 orders of magnitude increase in CFU on R2A after 7 days compared to the control). All mixed cultures (except for the one cultivated on glucose) consisted mainly of four different colony types similar to those of the batch-end sample

TABLE 1. Colony counts of bioreactor (for tar oil degradation) samples grown on various media.

Sample	Total Cell Count Per mL		
	R2A Agar	Diesel Oil Agar	Hydrocarbon Agar
Inoculum	$7.6 * 10^6$	$7.2 * 10^6$	$6.3 * 10^6$
Batch (after 16 h)	$1.1 * 10^8$	$1.2 * 10^8$	$9.1 * 10^7$
Batch (after 7 d)	$1.5 * 10^8$	$1.3 * 10^8$	$4.0 * 10^6$

in the bioreactor. These four colony types corresponded to the four identified species, while the other colonies (< 1 mm ϕ, slowly growing, < 10% of total CFU) corresponded to the unidentified organisms (< 5 positive results in physiological tests for identification due to their slow growth).

PAH Elimination by Bacterial Isolates

Only four isolates (nos. 6 to 9, all identified as *Pseudomonas aeruginosa*) eliminated significant quantities of the added PAHs (see Figure 1). Moreover, one bacterial mix (M3: equally distributed mixture of the batch-end isolates) showed an elimination of the tested PAHs. No isolates or mixtures showed PAH transformation within 3 days, except for isolate 6.

TABLE 2. Total number and diversity of CFU in the inoculum sample after precultivation on different substrates for 7 days.

Precultivation Substrate	CFU (R2A-Agar)	Number of Colony Types and Distribution
acenaphthene	$0.9 - 1.1 * 10^8$	four, equally distributed[a]
phenanthrene	$2.8 - 2.9 * 10^7$	four, equally distributed[a]
pyrene	$1.7 - 1.9 * 10^7$	four, equally distributed[a]
benz(k)fluoranthene	$1.4 - 1.5 * 10^7$	four, equally distributed[a]
glucose	$8.6 - 10.6 * 10^8$	four, small white predominant[b]
no substrate	$1.8 - 1.9 * 10^7$	four, equally distributed[a]

(a) Each of the four colony types was present at 10 to 50% total CFU on each of 4 agar plates (2 parallels, 2 dilution steps).
(b) More than 50% of total CFU.

PAH Elimination by the Mixed Culture
after Precultivation on Different Substrates

Figure 2 shows PAH elimination by the mixed culture after precultivation on various PAHs or glucose. In contrast to the isolates, this mixed culture showed an elimination (to a different extent) of all tested PAHs, regardless of precultivation. Although precultivation did not influence phenanthrene elimination, the elimination of anthracene and acenaphthene increased after cultivating the culture on acenaphthene or phenanthrene. Anthracene detection in the benz(k)fluoranthene (bkf)-precultivated sample could not be performed because of overlapping bkf-peaks (due to bkf-contamination of the sample). During phenanthrene degradation by different bacterial isolates or mixed cultures in former investigations, the formation of hydroxy naphthalene acid has often been observed. In our tests, this metabolite occurred only in small quantities (max. 60% of the degraded phenanthrene measured by peak height) compared to metabolite formation by other tested phenanthrene eliminating cultures (max. 120% in former tests with a bacterial mixed culture consisting of two different species of *Pseudomonas*, which were cultivated from petroleum-contaminated soil). In all tests, partial or complete elimination of this metabolite could be observed.

DISCUSSION

This investigation highlights two features of the tested mixed culture:

- The bacterial mixture showed rapid degradation of selected PAHs even under different substrate conditions.
- The composition of the mixed culture did not change after cultivating the sample on different substrates.

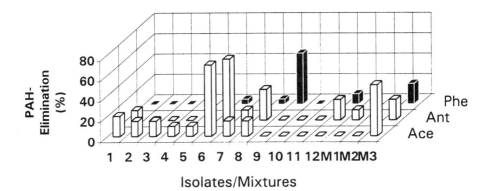

FIGURE 1. Elimination of selected PAHs (ace = acenaphthene, ant = anthracene, phe = phenanthrene) in test tubes by bacterial isolates (1-12) and mixtures (M1-M3).

132

Monitoring and Verification of Bioremediation

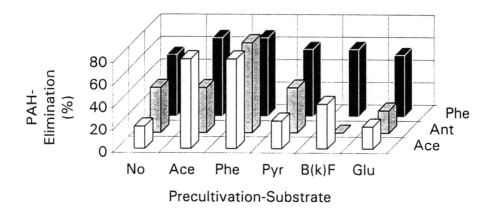

FIGURE 2. Elimination of selected PAHs (ace = acenaphthene, ant = anthracene, phe = phenanthrene) in test tubes by a mixed culture after precultivation on different substrates (no = no additional substrate, pyr = pyrene,
b(k)f = benz(k)fluoranthene, glu = glucose).

Compared with other mixed cultures (from bioreactors or contaminated soils)
in former investigations (Maue 1994), the mixed culture in this study was highly
stable. Besides the degradation capacity for selected compounds, this is an
important criterion for the use of such a mixed culture in bioremediation processes. An explanation for the high tolerance to changing environmental conditions could be the special sampling technique (Dreyer et al. 1995), in which
selected soils from the periphery of a tar contamination were taken, and the
cultivation of the well-adapted bacterial mixture (Kästner et al. 1991). The most
effective strain for PAH-degradation in pure culture was *Pseudomonas aeruginosa*,
but on the whole, the degradation potential of the isolates was small compared
to that of the mixed culture. This might be due to the symbiotic effects in the
mixed culture, that often are observed (Weißenfels 1990).

REFERENCES

Cerniglia, C. E. 1992. "Biodegradation of Polycyclic Aromatic Hydrocarbons." *Biodegradation*
3: 351-368
Dreyer, G., J. König, and M. Ringpfeil. 1995. "Polycyclic Aromatic Hydrocarbon Biodegradation
by a Mixed Bacterial Culture." In R. E. Hinchee, R. E. Hoeppel, and D. B. Anderson (Eds.),
Bioremediation of Recalcitrant Organics. Battelle Press, Columbus, OH. pp. 9-15.
Heitkamp, M. A., W. Franklin, and C. E. Cerniglia. 1988. "Microbial Metabolism of PAH:
Isolation and Characterization of a Pyrene-Degrading Bacterium." *Appl. Environ. Microbiol.*
54: 2549-2555.
Kämpfer, P., M. Steiof, and W. Dott. 1991. "Microbiological Characterization of a Fuel-Oil
Contaminated Site Including Numerical Identification of Heterotrophic Water and Soil
Bacteria." *Microb. Ecol.* 21: 227-251.

Kästner, M., M. Breuer, and B. Mahro. 1991. "Bakterienisolate aus unterschiedlichen Altlaststandorten zeigen ein vergleichbares Abbauprofil für PAK-und Ölkomponenten." *GWF Wasser-Abwasser 132*: 253-255.

Kiyohara, H., N. Takizawa, and K. Yana. 1982. "Rapid Screen for Bacteria Degradation of Water-Insoluble, Solid Hydrocarbons on Agar Plates." *Appl. Environ. Microbiol. 43*: 454-457.

Maue, G. 1994. "Aktivität und Abbauleistung bakterieller Biozönosen beim PAK-Umsatz." In: W. Dott and H. Rüden (Eds.), *Hygiene Berlin 18*.

Maue, G., P. Kämpfer, and W. Dott. 1994a. "Detection of PAH Degrading Bacterial Isolates Using a Rapid Fluorometric Method." *J. Microbiol. Methods 19*: 189-196.

Maue, G., P. Kämpfer, and W. Dott. 1994b. "Diversity of PAH Degrading Bacteria in an Airlift-Suspension Reactor System for Waste-Water Cleaning." *Acta Biotechnol. 14*: 337-345.

Steiof, M. 1993. "Biologische in-situ Sanierung eines mit Dieselöl kontaminierten Aquifers." In: W. Dott and H. Rüden (Eds.), *Hygiene Berlin 14*.

Weißenfels, W. D. 1990. "Mikrobieller Abbau von polycyclischen aromatischen Kohlenwasserstoffen (PAK) in definierten Submerskulturen und kontaminierten Böden." Dissertation at the Fachbereich Biologie of the Westfälische Wilhelms-Universität Münster, Germany.

Evaluation and Cost Assessment for Residual Military Pollution in Germany

Dieter Weth, Wolfgang Schröder, and Josef Korr

ABSTRACT

In the course of reunification of Germany, the corresponding dissolution of the National People's Army (NPA), and the withdrawal of military forces, large areas in the new German territory now have new uses. Problems have often been encountered due to contamination of soil, surface water, and groundwater from previous military operations in these areas. In this context, investigations were conducted to determine which sites are suspected of contamination, their potential toxic level, and their effect on health, and to evaluate the costs necessary to eliminate such historical pollution. These cost factors also play an important part in decision making concerning future land use. A data bank permits management, evaluation, and clearing of all the information flowing into it.

INTRODUCTION

The Federal Army of Germany, along with the other military forces, possesses land holdings encompassing approximately 453,000 ha. The land holdings of the former NPA in the eastern part of Germany encompass approximately 240,000 ha. Those of the Soviet army are around 256,000 ha. Due to the integration of the army of the former German Democratic Republic (GDR) into the Federal Army of Germany, the withdrawal of foreign forces from German territory, and the reduction of the total number of troops, the future use of these sites must be considered

The Head Departments of Finance (OFDs) in every state in Germany are responsible for all construction on sites owned by the Federal Republic of Germany, including the sites owned by the Federal Army. The OFD in Hannover, Lower Saxony, was appointed by the German Ministry of Defense and the German Ministry for Building and Construction to act as the supervising office. A task force unit (Project Management) was formed consisting of members of the OFD and experts from the consulting firm Prof. Mull and Partner.

INVESTIGATION STRATEGY

Historical pollution was investigated in accordance with recommendations of the Expert Committee for Environmental Matters in the German Parliament. In this multistage procedure, it is possible for decision-making processes to take place between individual consecutive stages of negotiation that permit assessment of decisions for further action with respect to individual areas on the basis of information obtained at each stage of the procedure. Investigations are normally split into three phases:

Phase I: Detection and initial assessment
Phase II: Risk assessment
     IIa: First technical investigations
     IIb: Detailed investigations
Phase III: Investigation, planning, execution, and monitoring of remediation

In the context of large-scale comprehensive historical pollution programs, a systematic survey of contaminated sites is required. For this purpose, a standard procedure and a standard, consistent information-gathering method are necessary to guarantee comparability of the results. These goals are achieved in the above procedure.

ELECTRONIC DATA PROCESSING

The project management of the OFD Hannover worked out a database solution that allows for management, evaluation, and clearing of all relevant information. With the goal of receiving a national model for all the basic data gathered, dialogues are held with the Federal Environmental Office in Berlin. This is an essential requirement for comparing the investigation results within the project as well as with other contaminated site examinations in Germany, and for processing the data for evaluation with formalized risk assessment models and cost estimation models.

Computer programs are at the disposal of the consulting engineering firms that are carrying out the reconnaissance study on site. Data collection programs EFA I and EFA II allow for storage of all the data and results in a uniform way. Parallel to that, all the data are stored at the existing database of the project management (Figure 1).

The interface of the database, INSA, provides user-friendly input and output of relevant data. INSA is designed to meet the information needs for which the building and planning administration and the relevant ministries ask.

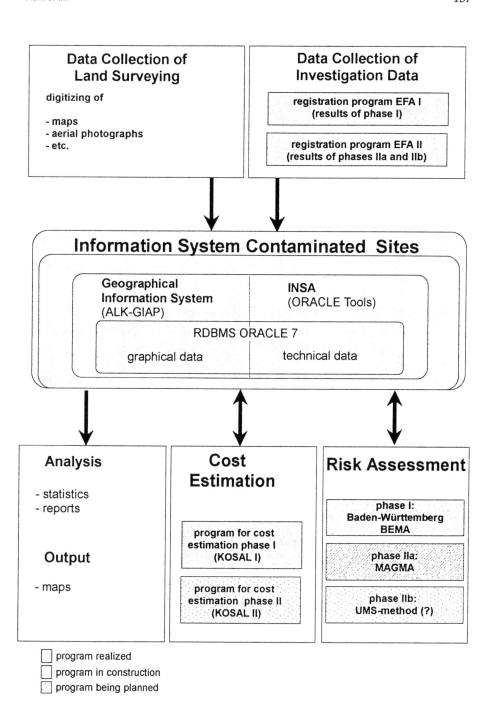

FIGURE 1. Programs for data registration and data processing.

Figure 2 is an overall view of the types of data handled. Both data for administrative purposes (e.g., project data) and specialized technical data are stored. Additionally, parallel to the capture of the administrative and technical data, registration of geographical data has begun by adding digitized maps, sketch maps, and aerial photographs to the database.

FORMALIZED RISK EVALUATION

Before it is possible to assess costs, it is necessary to evaluate the risk to affected subjects. The risk evaluation will provide the basis for the required remediation. An approximation of realistic costs can be achieved by taking into account the following influencing factors:

- Inventory of toxins
- Release conditions
- Regional background values
- Migration paths and conditions
- Affected subjects for protection/applications.

It is possible to determine the required course of action on the basis of these influencing factors. At the same time, on the basis of the knowledge gained in the process, it is possible to determine the required cleanup.

For the initial assessment after the Phase I model BEMA is used (BEMA was adapted from a model developed in the state of Baden-Württemberg), the data (e.g., measurements of pollutants in soil or groundwater samples, or soil characteristics) gained in Phase II (risk assessment), are used to determine risk with the help of the new model, MAGMA, developed by the German Environmental Ministry and the Federal Environmental Office in Berlin.

COST ESTIMATES

After the need for remediation and cleanup of suspected or confirmed contaminated sites has been ascertained by risk evaluation, a preliminary choice of steps to be taken will be made based on toxin inventory, geologic and hydrogeologic conditions, application, and the corresponding subjects for protection.

The possible remediation methods defined by preliminary selection will then be investigated according to the differentiated goals that are indicated. The following criteria will be investigated:

1. Quantities of contaminants for remediation or cleanup will be determined.
2. Technical feasibility (including space requirements/infrastructure, etc.) will be investigated.

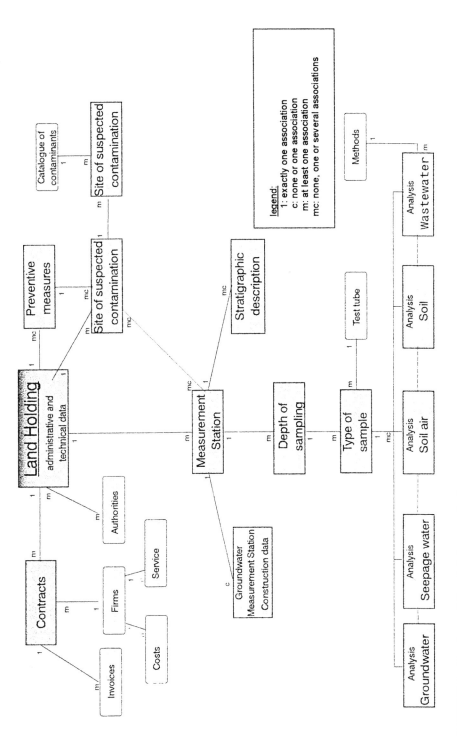

FIGURE 2. Design of the database solution.

3. Costs using the EPD-supported model KOSAL (developed by the German Ministry of Environmental in accordance with the Federal Environmental Office in Berlin and the OFD Hannover) will be calculated.
4. After Phase I, if large numbers of suspected contaminated sites with sources of the same kind are available, application of reduction factors will be statistically determined.
5. The economic/ecological effectiveness will be studied.

CONCLUSIONS

The OFDs in every state in Germany are investigating approximately 1,000 land holdings with a former military use where contaminants have spread predominately into soil, but also into groundwater. The further use of these sites depends on the extent, the intensity, and the risk for the biosphere. This risk must be assessed in more detail.

To manage, administer, and evaluate the captured data, a database system has been developed that is regularly further extended. The system now includes programs to register, analyze, and evaluate data. In the context of data evaluation, risk assessments and cost evaluations, which are supported by data processing, are executed. A database user interface has been drawn up to provide exact information concerning the registrated data. In addition, a geographic information system is being developed to depict results graphically.

REFERENCE

Heiwolt, G., D. Weth, and W. Schröder. 1993. "Datenbanksystem für das Projektmanagement des Altlastenprogramms Ost der Bundeswehr."

A Bench-Scale Bioreactor to Bioremediate Dredged Sludge and Soil

Mark Carpels, Luc Kinnaer, Diana Vanhoutven,
Helmut Elslander, Sandra VanRoy,
Liliane Hooybergs, and Ludo Diels

ABSTRACT

Detecting the microbiological degradation of pollutants on a laboratory scale requires vast amounts of agar and petri dishes. Translating results into a working industrial process is still a matter of trial and error. The bench-scale test system described here meets the need for a larger scale, completely closed and fully controlled instrument that can be used to design full-scale bioreactor installations.

INTRODUCTION

Some aspects of microbiological behavior require the use of bench-scale reactors instead of laboratory growth tests to determine the feasibility for industrial application. In most cases, when running laboratory tests, sampling frequency must be limited because of the small amount of test material. When a good characterization has already been made this is not a problem, but when the material being studied is not well known, interpretation errors can arise.

For industrial applications, growth rate is one of the most important parameters to identify. Microorganisms that show a 100% breakdown efficiency but no significant growth have no industrial relevance. Finding growth or breakdown rates with only a few experimental points also can lead to serious errors. Linear exponential or Monod models may give the same initial and end results, but with completely different kinetics. In addition, the overall breakdown rate can easily be 5 to 10 times lower in the field than under laboratory conditions because laboratory temperatures can range from 20 to 25°C, but in industrial applications equivalent specific inputs may never be economically possible. Another important factor is the evaporation of the contaminant during aeration. At the laboratory scale it is important to have a control over aeration, because removal of oil by aeration can be misinterpreted as biological activity.

Experiments in 250-mL Erlenmeyer flasks show initial oil removal rates of 500 mg/L/d, due only to evaporation, without any microbiological activity.

This research was carried out under the sponsorship of VLIM (Flemish Impulse Programme for Environmental Technology) to build a bioreactor system capable of remediating oil-contaminated gasoline stations. Two problems had to be overcome: (1) small-scale bioreactors were required because the installation had to be mobile, and (2) remediation had to be faster than traditional soil vapor extraction technology.

REACTOR DESIGN

Three identical stainless steel, closed, batch test reactors were built, each with a total volume of 200 L. Soil volume can be up to 100 L to provide enough material to ensure noninteracting sampling. A computer equipped with the Labview program controls a number of functions and fulfills data acquisition.

The internal temperature of the reactor is proportional-integral-differential (PID)-controlled (constant or with variations such as day-night cycles) through a double-wall water circuit. The temperature difference between the wall and the bulk cannot exceed a certain amount, to prevent microbial death. The inlet gasflow rate (oxygen, nitrogen, air) can be set or controlled with the carbon dioxide, oxygen, or hydrocarbon concentration. The outlet of the gas can be automatically sampled and analyzed for hydrocarbons by gas chromatography (GC), and for oxygen and carbon dioxide content. The data are fed into the computer. The solid phase can be sampled so that the reactor is opened only for a few seconds to ensure minimal interaction with the outside.

Nutrients, liquids, extra inocula, etc. can be automatically injected without perturbation of the system. The soil is mixed using special abrasion-free paddles, permitting tests with dry, sandy material.

RESULTS

The optimal temperature was tested in smaller-scale reactors and was set at 28°C. Nutrients were added, but concentrations have not yet been optimized. An inoculum was used that gave the best results compared with about 20 other selected microorganisms in 3-L reactors. The inoculum was composed of *Acinetobacter* sp. LH 168 or *Sphingomonas* sp. LH214. The initial oil contamination was 6,000 mg/L, and the reactor was filled with a 50-L soil suspension inoculated to an initial bacterial concentration of 10^6 colony-forming units (cfu)/g soil.

Table 1 shows the test conditions on which the measurements were carried out. Gas and soil were sampled twice each day. The experiment took 5 days. Gas hydrocarbons, soil hydrocarbons, oil degraders, microorganisms, and pH were traced. Figure 1 shows an example of those results. From those results, it

TABLE 1. Test run conditions. Different soil dry matter, gasflow rates, and mixing intensities were combined in the reactor.

Test No.	Dry Matter %	Gasflow Rate (L/h)	Frequency (Hz)
4	80	16	20
5	80	16	15
6	80	16	15
7	50	60	15
9	50	100	15
11	70	600	20
12	70	600	20
14	70	300	20
15	70	300	20
16	70	600	30
17	70	600	15
18	70	300	30
19	70	300	15
20	70	100	15
21	70	200	15

was possible to calculate maximal degradation velocities and the end point oil content.

As seen in Figure 2 the maximum growth velocity became optimal (0.08 h^{-1}) with a dry matter content of 70% and a mixing frequency of 15 Hz, independent of gasflow rate (two experiments with dry matter content of 50% and 80% are shown too). This means that gasflow rates can be less than 100 L/hr. The same picture is found with the end point oil content where, under conditions of 15-Hz mixing frequency and 70% dry matter, about 100 mg oil/kg soil was found. With higher and lower dry matter content, higher end points where found.

Surprisingly, the mixing efficiency did not enhance degradation efficiency. This can be seen in Figure 3. The experiments with a double mixing rate showed a significantly lower μmax and a markedly higher ($>1{,}000$ mg/L) oil content at the endpoint.

The results are being used in the scale up of the installation. Experiments are being carried out with lower mixing rates and lower temperatures. The

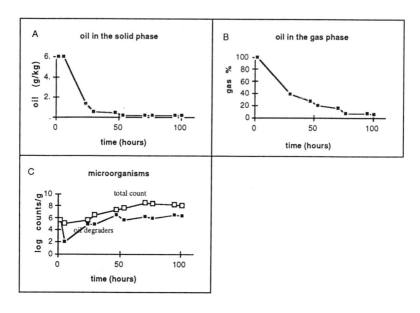

FIGURE 1. Results of a bioremediation test run. The conditions of test run 17 (see Table 1) were used to biodegrade soil contaminated with 6 g mineral oil/kg. The biodegradation of oil in the soil phase (A) and gas phase (B) are presented as functions of time. The initial oil concentration in the gas phase (=0.15 g oil/L) was presented as 100%. Total cfu and the mineral-oil degraders cfu are also presented as functions of time (C).

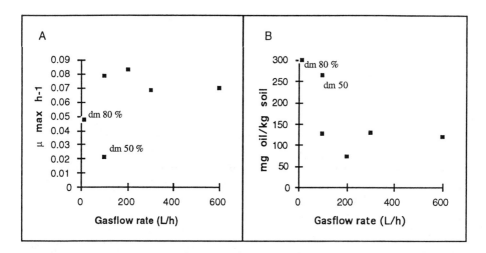

FIGURE 2. Influence of dry matter on growth parameters. Maximal growth rate (μmax) (A) and final mineral oil concentration (B) are presented as a function of the gasflow rate. Dry matter (dm) was 70% unless otherwise indicated. The mixing rate was 15 Hz.

reactors are being evaluated to determine the possibility of using a consortium of microorganisms to degrade polycyclic aromatic hydrocarbons (PAHs) in dredged sludge with heavy metal cocontamination. The experiment was conducted with 80 L of in situ density sludge. Temperature was set at 28°C, mixing rate at 15 Hz, and gasflow rate at 100 L/h. Figure 4 shows the PAH analysis

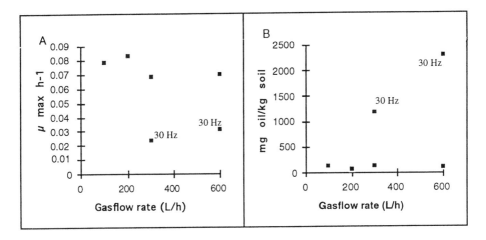

FIGURE 3. Influence of mixing rate on growth parameters. Influence of mixing rate was presented by showing the maximal growth rate (A) and final mineral oil concentration (B) as functions of the gasflow rate. The experiments were done with a pulp density of 70% and a mixing rate of 15 Hz unless otherwise indicated.

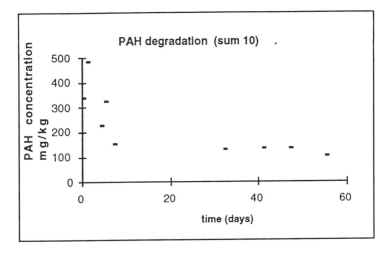

FIGURE 4. PAH degradation in dredged sludge. PAH concentration is presented as a function of time. The dry matter content is 50%.

of the soil samples. Heavy-metal-resistant PAH degraders will be tested to see if lower end points can be achieved.

DISCUSSION

The use of fully automated and completely controlled bench-scale reactors made it possible to establish microbiological degradation activities in correlation with some technological process parameters. This led to an innovative degradation process for oil-contaminated soil.

Composting Oily Sludges: Characterizing Microflora Using Randomly Amplified Polymorphic DNA

Anders Persson, Mikael Quednau, and Siv Ahrné

ABSTRACT

Laboratory-scale composts (total volume 100 L), in which oily sludge was composted under mesophilic conditions with amendments such as peat, bark, and fresh or decomposed horse manure, were studied with respect to basic parameters such as oil degradation, respirometry, and bacterial numbers. Further, an attempt was made to characterize a part of the bacterial flora using randomly amplified polymorphic DNA (RAPD). The compost based on decomposed horse manure showed the greatest reduction of oil (85%). Comparison with a killed control indicated that microbial degradation actually had occurred. However, a substantial part of the oil was stabilized rather than totally broken down. Volatiles, on the contrary, accounted for a rather small percentage (5%) of the observed reduction. RAPD indicated that a selection had taken place and that the dominating microbial flora during the active degradation of oil were not the same as the ones dominating the different basic materials. The stabilized compost, on the other hand, had bacterial flora with similarities to the ones found in peat and bark.

INTRODUCTION

Composting has been suggested as a possible technique for local, final treatment before disposal of oily residues obtained from separation processes for sludges from oil separators, etc. This technique is performed at a handful of treatment stations in Sweden. Outdoor, semifull-scale composts have indicated a rapid degradation of the oil content. Little knowledge exists, however, on the mechanisms behind the observed reduction. This project, initiated by the Swedish Association of Solid Waste Management and the Biotechnology Research Foundation, is a "down-scaled" process, providing controlled or measurable conditions over time.

To characterize the microflora that may be responsible for the degradation, we combined basic studies such as product (oil) reduction, bacterial activity (respirometry), and bacterial numbers in the compost process with a genomic "fingerprint" of the isolated bacteria, i.e., RAPD, a polymerase chain reaction (PCR)-based method. The standard PCR consists of repeated cycles of template DNA denaturation, primer annealing, and primer extension. By reducing the stringency of the primer annealing step, a single primer that has no known homology to a genome can anneal at sites for which the match is imperfect and allow certain fragments of the genome to be amplified. Products, after being resolved by electrophoresis, give rise to fingerprints of the genome (Welsh and McClelland 1990; Williams et al. 1990). A typical RAPD gel is shown in Figure 1.

EXPERIMENTAL PROCEDURES AND MATERIALS

Composting Equipment and Maintenance

The compost vessels consisted of insulated, cylindrical galvanized sheet-iron barrels of 100 L total volume (inside diameter 400 mm, height 880 mm).

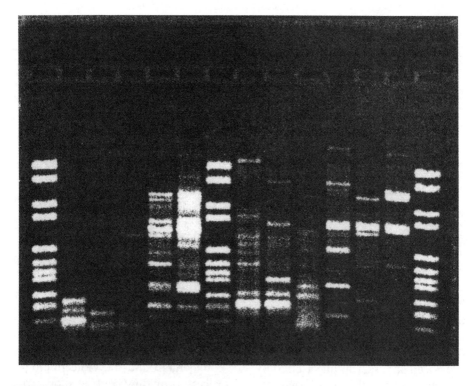

FIGURE 1. A typical RAPD gel. Lanes 1, 7, and 14 are size standards used for normalization of the gel.

The working volume was approximately 75 L. Air was introduced through a perforated plate in the bottom and the cover was equipped with a butyl rubber cork for gas sampling, a thermometer, and an air exit to a hydrocarbon trap (activated carbon). Airflow was measured at the exit, and was maintained between 10 and 20 L/h, yielding aerobic conditions (no CH_4 detected) with CO_2 levels between 2 and 5%.

The compost was mixed weekly by rolling the barrels. Dry weight was kept at approximately 50 to 60% (weight) by intermittent addition of water. Temperature, weight, and pH were not controlled, only measured. Sampling was performed as the composts were mixed.

Four different composts were studied, based on oily sludge and (1) peat; (2) peat, bark and fresh horse manure (mixed with straw); (3) peat, bark, and decomposed horse manure (in straw); and (4) a killed control similar to (1), with 2% $HgCl_2$ added. The control was, due to its environmental hazard, performed at a smaller scale, in a 1-L temperature-controlled (30°C) vessel. Mixing proportions are given in Table 1.

Analyses

Chemical Analyses. The hydrocarbon (total extractable oil, TEO) content was measured according to Swedish Standard Method SS 02 81 45 (modified for soil samples). This method is based on infrared (IR) and gives the concentration according to a given standard (50 vol-% n-hexadecane and 50% isooctane). TEO was quantified as total and nonpolar fractions (the amount not adsorbed to aluminum oxide). Qualitative information was provided by gas chromatograph (GC) analyses.

Nitrogen was determined as total nitrogen (according to Kjeldahl) and phosphorous as phosphate. The pH was measured in overnight (airtight) samples of compost and water (1 + 1 by vol). Exit gas was analyzed by GC, providing information on the content of CO_2, CH_4, and H_2S.

Dry weight (DW) determinations were analyzed after 24-h drying at 105°C, ash content (and correspondingly volatile organics) as the fraction remaining (reduced) after heating the dried sample at 550°C (4 h).

Microbiological Analyses (Other Than Those Based on DNA). In situ respirometry was calculated from knowledge of CO_2 content in exit air, airflow, and compost weight.

Enumeration (colony forming units, CFU) of bacteria was made from serial dilutions on two different types of agar plates: (1) standard "total number" TGEA (tryptone glucose extract agar, Difco Laboratories), 3 days incubation at ambient temperature and (2) potential hydrocarbon degraders medium consisting of a hydrocarbon mixture (2 to 5 g/L diesel), trace element solution, phosphorus (KH_2PO_4, 0.2 g/L), nitrogen (NH_4Cl, 1 g/L), and Bacto agar (15 g/L, Difco Laboratories), pH 7.0, 7 to 8 days incubation at ambient temperature.

Isolates for further study were picked from these plates; restreaked twice onto TGEA plates; and checked for purity, morphology, and mobility by

TABLE 1. Configuration, initial conditions, and TEO reductions in composts (1) through (4).

Compost	Configuration, % (by weight)					Initial conditions (g/kg DW)				Residual TEO	
	Oily sludge	Peat	Bark	Fresh horse manure	Decomposed horse manure	pH	TEO	N	P	g/kg DW	(% reduction)
(1)	53	47	—	—	—	5.5	42	3.8	0.6	17	(55)
(2)	43	19	19	19	—	6.5	29	6.1	2.0	11	(70)
(3)	53	23	12	—	12	6.0	45	4.9	1.1	7	(85)
(4)[a]	52	46	—	—	—	5.5	80	—	—	60	(25)

(a) 2% $HgCl_2$

microscopic examination. Verification of hydrocarbon-degrading potential was scored as growth in 5 mL liquid medium (2), agar excluded. Gram reaction and oxidase tests followed standard procedures (Cerny 1976, 1978; Kovacs 1956).

Randomly Amplified Polymorphic DNA. A 1-mL overnight culture was centrifuged, washed twice in 1,000 μL of sterile distilled H_2O (SAQ), and resuspended in 250 μL of SAQ. The cells were disintegrated by vigorous shaking together with 10 glass beads (0.2 mm in diameter) by an Eppendorf Mixer 5432 for 1 h at 4°C. The disrupted cells were pelleted by centrifugation, and the supernatant was used in the PCR reaction.

The PCR reactions were carried out in a Perkin Elmer Cetus apparatus. Crude cell extract (1 μL) was used, and amplification reactions were performed in volumes of 50 μL using 0.5 U of Taq-polymerase (Boehringer Mannheim) and the buffer provided by the manufacturer, including 1.5 mM $MgCl_2$. The nucleotide concentrations were 0.2 mM of each dNTP, and the primer concentration was 12 μM. The primer 5´- CCG GCG GCG - 3´ was synthesized by Symbicon AB, Umeå, Sweden. The reaction was overlaid with oil and cycled through the following temperature profile: 94°C, 45 s; 30°C, 2 min; 72°C, 1 min for four cycles followed by: 94°C, 5 s; 36°C, 30 s; 72°C, 30 s for 26 cycles (the extension step being increased by 1 s for every new cycle). The PCR session was terminated with 75°C for 10 min.

PCR reactions (20 μL) were run on agarose gels (1.5% in Tris-Borate buffer) at 100 V for 2.5 h. Gels were stained with ethidium bromide, and photos were taken under ultraviolet (UV) light.

Negatives were scanned into a computer, using an Ultrascan XL densitometer (Pharmacia, Bromma, Sweden), and all the subsequent steps (normalization of the gels, creation of lists of gel tracks to compare, and calculation and printing of dendrograms) were made using the GelCompar 3.0 software package (Applied Maths, Kortrijk, Belgium). The gel tracks were compared using the Pearson Product Moment Correlation Coefficient (r), and the dendrograms were calculated using the algorithm of Ward (1963).

RESULTS AND DISCUSSION

The results presented are, unless otherwise stated, from compost (3), showing the largest reduction in TEO.

Total Extractable Oil

The oil in the sludge consisted of two fractions, one with a GC resembling lubricating oil and one resembling light fuel or diesel oil (data not shown). Table 1 shows the composition of the four different composts studied. Peat was used as a medium to evenly homogenize and distribute the oily sludge. Composts (2) and (3) were supplemented with nutrients (manure), and (2) also with an extra (besides the oil) and more easily degradable carbon source because it contained fresh manure. Compost (3), with decomposed horse manure added, showed the largest reduction of TEO, from 45 to 7 g TEO/g DW (85%). In contrast, and possibly depending on nutrient limitations, reduction in peat-based compost (1) was from 42 to 17 g TEO/g DW (55%).

The killed control (4) indicated that microbial degradation actually had taken place. A substantial part of the minor reduction in TEO (75%) could be explained by evaporation, because it was detected in the hydrocarbon trap. Results from (1) through (3) showed that approximately 5% of the observed reductions in these active composts were due to evaporation during the first part of the process. GC analyses indicated the presence of hydrocarbons up to a corresponding $n\text{-}C_{14}$ in the exit air (data not shown). The relatively low evaporation is probably due to the prevailing mesophilic conditions.

Microbial Activity

Temperature. The compost processes were carried out under mesophilic conditions, and the temperature ranged from 20 to 35°C. Figure 2 shows the temperature profile in compost (3). Higher temperatures during the first part corresponded to higher carbon dioxide production (the elevated temperature at 130 days was due to external heating, which did not increase the microbial activity, see respirometry below).

Respirometry and Bacterial Numbers. As seen in Figure 2, carbon dioxide production follows oil degradation (short time variations are probably caused by difficulties in determining the airflow rate). However, assuming that all produced CO_2 was detected in the exit air (pH was in the range of 5.5 to 6.5,

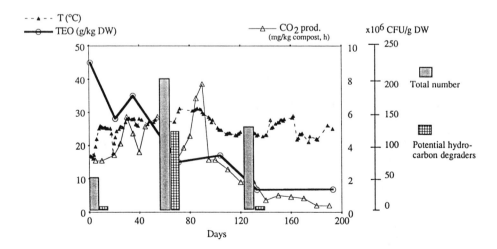

FIGURE 2. Temperature, TEO, CO_2 production, and bacterial numbers (CFU) of compost (3). Isolates designated X: 75: Z in Figure 3 correspond to samples taken after 75 days, X: 125: Z to 125 days.

see Table 1), a substantial part of the oil (corresponding to approximately 40%) was only partly degraded. Addition of extra nutrients or elevated temperature (130 days, Figure 2) did not further increase the degradation or the activity, indicating that the remaining oil consisted of (stabilized) nondegradable residues.

CFU, as expected, showed the highest numbers (both as total number and as potential hydrocarbon degraders) during the most active period (Figure 2). The source materials showed total numbers between 10^7 (sludge, peat, and bark) and 10^9 (manure) CFU/g DW.

Characterization of Isolates

Phenotypes. Phenotypic and RAPD studies are summarized in Figure 3. About 80% of the isolates on agar plates collected as "potential" hydrocarbon degraders were verified as actual degraders. Furthermore (and not surprisingly), several total number isolates showed this potential (about 40%). It should be kept in mind that the study points out any isolate capable of visible growth on a hydrocarbon mixture (diesel).

Most of collected isolates, and verified hydrocarbon degraders, could be described as gram-negative, oxidase-positive rods, some motile. The collected isolates may represent only a minor fraction of the microorganisms present in the compost, and the lack of gram-positives could be due to the cultivation conditions for the isolated bacteria (gram-positives, though, were found in great numbers in the fresh horse dung, data not shown). Fungi also could have been present in significant numbers.

RAPD. RAPD gives a fingerprint, and the dendrogram in Figure 3 shows how similar these prints are. The degree of similarity is expressed as a horizontal distance. There is a clustering of isolates from different basic materials: sludge in cluster 3, bark and peat in no. 4 (and 1), and manure in no. 7. It is not always possible to distinguish between isolates from different periods of the compost process (indicated X:..**75** or **125**:Z). However, most of the isolates taken from the period when the TEO degradation rate was highest (X:..**75**:Z) are found in cluster 6 (3 out of 4 isolates). Isolates from the end of the process (X:..**125**:Z) are more distributed and found in clusters 1, 2, 5 (3 out of 8), and 7. Cluster no. 5 contains a mixture of isolates collected at the end of the compost process and from bark, manure, and peat.

The clustering is a consequence of the different environmental conditions found, and it seems as if the dominating flora during the degradation are not the same as those dominating in the basic materials. A selection has taken place, and, as the material is stabilized during the compost process, the flora turn into a mixture with similarities to the flora found in the original ("stabilized") materials (clusters 1, 5, and 7).

Verified hydrocarbon degraders are frequently found among sludge isolates in cluster 3. Accordingly, these isolates probably are not responsible for the TEO degradation in the compost. These are, on the other hand, more likely

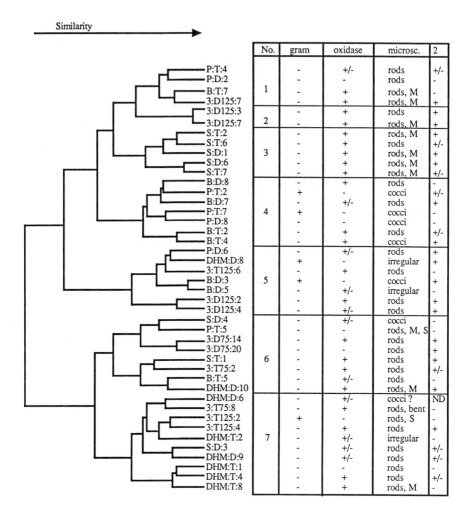

FIGURE 3. Dendrogram and characteristics for isolates from compost 3 (3). Isolate names: X:Y:Z—X = Source (P = Peat, B = Bark, S = Sludge, DHM = Decomposed Horse Manure, Y = Medium (agar plates) (D = Diesel, T = TGEA), and sampling event, Z = isolate number, Columns: 2 = growth in liquid diesel medium, ND = Not Determined.

those found in clusters representing isolates from the compost process, that is nos. 5 and 6. It would have been interesting to increase the number of isolates during the active period of the compost process to be able to follow the shift in the microbial flora more closely.

Studying a diversified flora one should, together with RAPD, run a small number of fast and simple phenotypic tests such as gram-reaction, catalase, and oxidase tests. In this way, the somewhat confusing dendrograms caused by meaningless comparisons between groups of bacteria that are not related, e.g., gram-positive and gram-negative organisms, can be avoided. An optimization of the primer for the type of isolates used, and possibly the use of several primers, may give better results. Also, because RAPD is a low stringency method, the reproducibility has to be further examined.

REFERENCES

Cerny, G. 1976. "Method for the distinction of gram-negative from gram-positive bacteria." *European J. Appl. Microbiol.* 3:223-225.

Cerny, G. 1978. "Studies on the aminopeptidase test for the distinction of gram-negative bacteria from gram-positive bacteria." *European J. Appl. Microbiol.* 5:113-122.

Kovacs, N. 1956. "Identification of *Pseudomonas pyocyanea* by the oxidase reaction." *Nature.* 178:703.

Ward, J. H. 1963. "Hierarchal grouping to optimize an objective function." *J. Am. Statist. Assoc.* 58:336-344.

Welsh, J., and M. McClelland. 1990. "Fingerprinting genomes using PCR with arbitrary primers." *Nucleic Acid Research.* 18:7213-7218.

Williams, G., A. Kubelik, K. Livak, A. Rafalski, and S. Tingey. 1990. "DNA polymorphisms amplified by arbitrary primers are useful genetic markers." *Nucleic Acid Research.* 18:6531-6535.

Hierarchy of Treatability Studies for Assured Bioremediation Performance

Diane L. Saber

ABSTRACT

Three levels of treatability study testing (Levels I, II, and III) have been proposed as tools for evaluation of specific bioremediation programs. Due to inherent differences between bioremediation designs, contaminant/matrix complexity, program phase, intricacies of design, cleanup standards, and cost, specific treatability study approaches may vary. Selection of the appropriate treatability study level is dependent upon the needs of the program and the level of confidence required for selection of bioremediation technology for full-scale application. Testing at the appropriate level ensures maximum information gathering for the specific program phase without incurring unnecessary costs or schedule extensions. Level I testing (Proof of Concept) is a simple, relatively inexpensive bench-scale test conducted for basic qualification of the technology, without regard to specific design, during the feasibility study of the technology evaluation phase of the program. Level II testing (Engineering Evaluation), executed during the pre-design phase, is used to specify the system necessary for optimal performance given site conditions, cleanup standards, and client preferences. This phase may identify pretreatment and posttreatment processes, synergistic technologies, or preliminary design criteria. Level III testing (Integrated Predesign) entails the use of multiple disciplines to construct a clearer, more precise appraisal of all requirements and anticipated performance from the full-scale program using the assigned design or approach.

INTRODUCTION

Over the past 15 years, millions of dollars have been spent on qualifying particular technologies for site application through the use of treatability studies. Because there are no specific standards for these studies, results are often highly variable, inconclusive, or limited to one particular design. For example, identical bioremediation "pan studies" often are conducted for widely differing

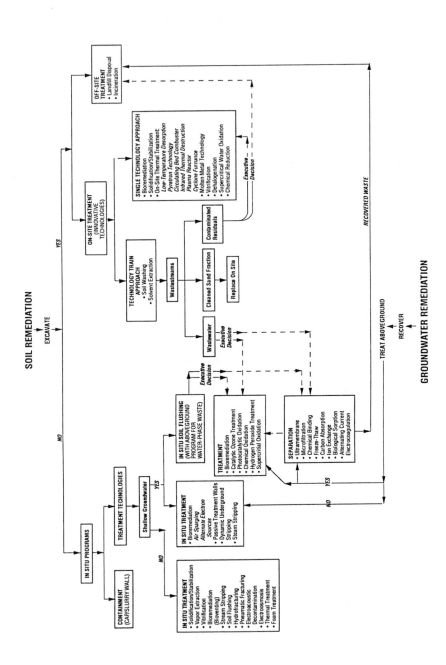

FIGURE 1. Remedial action decision tree.

sites, such as gasoline station spills and Superfund sites. Information gathered from routine testing methods may be highly applicable for one site but grossly inadequate for the other. Consequently, some treatability studies often are repeated to gather required information for technology selection or design completion. Alternatively, some treatability study testing yields too much information for the program level. The problem, therefore, is that generic treatability studies do not suffice for all types remediation sites.

A hierarchy of treatability study testing has been proposed here: Level I or Proof of Concept, Level II or Engineering Evaluation, and Level III or Integrated Predesign. The purpose of the proposed tiers is to reduce program redundancy, program costs, and treatability study costs, and to increase accuracy of final program costs and schedule requirements. In all, the tiered approach is used to improve program performance. The level of testing is dependent upon the following: program phase, proposed design, contaminant mix, matrix, complexity of the site, cleanup standards, and cost. A wide variety of cleanup technologies are available for in situ or aboveground treatment of soil and groundwater (Figure 1). Bioremediation technology encompasses at least five basic ex situ designs and three basic in situ designs (Figure 2). There can be variations on these basic designs. As such, the treatability study testing should seek to answer questions specific to the particular site for a particular phase of the remediation program and with particular goals in mind.

PROPOSED TREATABILITY STUDY LEVELS

Level I—Proof of Concept

The Level I Proof of Concept involves basic technology qualification. This level of testing is applicable when all technologies are under consideration and no design has been assigned. It is generally performed prior to or as part of a feasibility study evaluation using simple or complex matrices. The cost is minimal for Level I testing. Level I testing focuses on performance of the technology under optimal conditions and considers methods or accessory technologies that would enhance core technology performance. Considering bioremediation, removal of particular compounds through microbial action is central to Proof of Concept testing. Results are based upon the laboratory test design (generally simple, such as soil slurries or pan studies). Results from this level of testing will be incorporated into Level II or Level III testing, if necessary. Level I testing results generally do not yield effectiveness of the technology using alternative bioremediation designs, full-scale program costs, full-scale program length, or effectiveness of the technology for the entire site.

Example—Level I Testing. Soil and groundwater have been contaminated with pentachlorophenol, creosote, and chromium. No remedial design has been assigned (in situ or ex situ designs considered). Cleanup criteria have

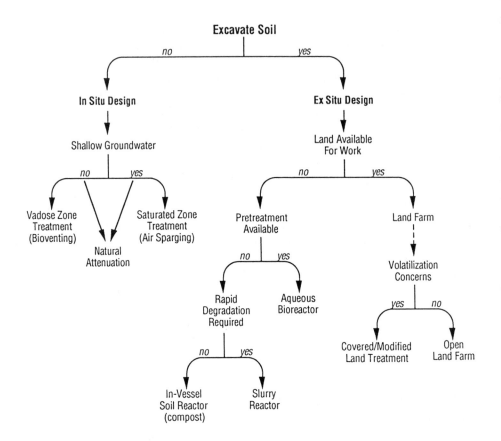

FIGURE 2. Bioremediation design decision tree.

been assigned for the site, and the treatability study is performed prior to the feasibility study phase of the project.

For this example, testing includes microbial enumerations of degradative strains and biodegradation tests in both soil and water. Microbial metabolism of all compounds of concern to the specified cleanup levels is tested using abiotic controls, rigorous analytical analysis, and sacrificed samples in triplicate. For this test, a 5% slurry was chosen as the soil test.

Results from the example indicate that pentachlorophenol and the creosote degraded to cleanup standards in water in 96 days. Pentachlorophenol and creosote also degraded to cleanup standards in soil in 4 weeks. Additionally, chromium levels affected biodegradation, but the metal was not removed.

Proof of Concept testing for this example derived the following results: bioremediation could achieve cleanup standards in both soil and water within discrete time periods (therefore, it would be a viable alternative for cleanup consideration), soil remediation was achieved using a 5% slurry solution (as tested, the soil will require excavation and slurry treatment), chromium was not removed (requires a metal-removal technology), and an active native microbial population exists at the site (site does not require microbial augmentation).

Level II—Engineering Evaluation

Level II or Engineering Evaluation seeks to answer questions from Level I testing. To this end, synergistic technologies are identified and evaluated. Engineering Evaluation testing is performed during predesign or after the feasibility study. Generally, the design has been selected (in situ or ex situ) and information from Level I testing (one or more technologies) is incorporated. Other considerations, such as implementability, agency/client concerns, program schedule, and cost are assessed. Level II testing provides costing information and pretreatment/posttreatment requirements, but limited information on program length. Additionally, results may be applicable only to discrete locations at the site (sample dependent). The cost associated with these studies is moderate.

Example—Level II Testing. Approximately 1,000 tons of soil have been contaminated with high levels of paraffinic total petroleum hydrocarbon (TPH) compounds. A design has been suggested, and no cleanup criteria have been assigned as yet. The client had a specific budget and is considering both in situ and ex situ treatment, whichever is least expensive. The testing was performed in the predesign phase of the project.

For this example, a general process flow diagram was constructed, integrating a variety of technologies: soil flushing, soil washing, biological soil slurry, and aerobic bioreactors. Following the proposed treatment-train sequence, the technologies were subjected to Level I testing, using soil or water from the previously tested technology. Each technology was evaluated for performance. Technologies were costed independently, based upon the quantity of soil affected, and the system was costed in total.

Results indicated that the soil flushing tests were not successful due to the permeability of the soil. Soil subjected to washing retained residual quantities of TPH. Biological soil slurries, performed on the washed soil, did not remove additional quantities of TPH from the soil. However, the aerobic bioreactors removed more than 95% of TPH from water retrieved from the soil washing process. Engineering Evaluation testing for this example, therefore, derived the following results: the soil may not be treated in place using soil flushing (ex situ treatment required), excavated soil subjected to soil washing and bioslurry would require additional treatment to remove residual quantities of TPH (an accessory technology is required after these treatments or a more aggressive technology should replace them), and TPH removed into the water phase from the soil washing process was degraded microbially (liquid-phase biotreatment is possible). Due to the inherent limitations of the system and the resulting costs associated with the additional remedial treatment required, landfilling the material was recommended.

Level III—Integrated Predesign

Level III or Integrated Predesign consolidates the results from multiple disciplines for a more complete analysis of all parameters involved with a full-scale program. Groundwater modeling, risk assessment data, soil testing, biological investigations, and examination of synergistic technologies may be included in Integrated Predesign testing. Heavy emphasis is placed on the amalgamation of pertinent data. Level III testing generally occurs after a specific remedial design has been approved, during the predesign phase. The site may consist of a complex matrix and contain a complex mixture of contaminants. Information from numerous Level I tests are applied to the specific remedial design (most often an in situ design). Other considerations such as maximum design performance, overall project length, program schedule, construction requirements, and cost are reviewed. Integrated Predesign testing provides detailed costing information for the full-scale project and elucidates all requirements for 100% design. Depending on site characterization and sampling, however, Level III testing may not prove the effectiveness of the program for the entire site. Integrated Predesign testing can be moderate to expensive in cost, but higher confidence in the results can be expected.

Example—Level III Testing. A 10-acre site, located within a wetlands area, was contaminated with a mix of organics (volatiles and semivolatiles), inorganics and construction debris. Soil and groundwater were affected. Cleanup criteria have been assigned. In situ treatment was preferred. The Level III study was performed as part of the predesign phase (25% design), concurrent with the risk assessment for the site.

Level III testing included Level I and Level II testing of five different technologies, with integration of risk assessment requirements. Two groundwater models were performed (BIOPLUME II and an oil recovery model) and results were incorporated into the Level III testing. The candidate design was selected, and designated technologies were subjected to an increased level of testing to meet the needs of the design (oxygen uptakes were recorded and degradation rates per organic species were performed). Technologies were tested in sequence, according to their place in the treatment train. All experiments sought to exactly mimic field conditions. Results from one test were integrated into the following test so that the full-scale system could be completely evaluated.

Integrated Predesign testing for this example, therefore, derived the following results: the proposed treatment-train design met cleanup criteria requirements if certain provisions were made ("Hot Spot" excavations were required); specific soil contaminants may be biotreated in situ, but groundwater would be treated in aboveground bioreactors and reinjected (a slurry wall is required); pretreatment with hot water flushing prior to aggressive biotreatment is possible (but cost is prohibitive); and some contaminants may remain in the soil, but are not of concern according to the risk assessment (the site may require capping and access restriction). Results also indicated that the program would be long, but limited excavation would be required. Cost was estimated for installation of the system and projected per year of operation.

CONCLUSIONS

A hierarchy of treatability study testing (Proof of Concept, Engineering Evaluation, and Integrated Predesign) has been proposed. Each level is designed to meet the needs of the program during a specific phase of work. Other factors such as bioremediation design, contaminant/matrix complexity, cleanup standards, and cost are influential on the level of treatability study testing (Table 1). The proposed hierarchy can help eliminate costs associated with (1) duplication of testing for additional information; (2) unnecessary testing for the level of design or program phase; (3) inaccurate costing of the full program, based on an incomplete data from testing; and (4) early identification of additional technologies or site manipulations required for the successful application of bioremediation technology. Although the focus has been on bioremediation technology in this paper, a wide variety of technologies may be evaluated using the proposed testing scheme.

TABLE 1. Comparison of treatability study levels.

Treatability Study Level	Phase of Remediation Program	Description	Contamination Level	Matrix	Cost[e]
LEVEL I: Proof of Concept	Feasibility Study	Basic Technology Qualifications	Simple[a] or Complex[b]	Simple[c] or Complex[d]	$8,000– 20,000
LEVEL II: Engineering Evaluation	Predesign	Cooperative Technologies Tested	Simple[a] or Complex[b]	Simple[c] or Complex[d]	$20,000– 80,000
LEVEL III: Integrated Predesign	Predesign	Multiple Disciplines Involved	Complex[b]	Complex[d]	$80,000– 5,000,000

(a) A matrix consisting of primarily one contaminant.
(b) A matrix consisting of a variety of contaminants, organic and/or inorganic.
(c) Good soil hydraulic conductivity (10^{-3} cm/s or more).
(d) Poor soil hydraulic conductivity (10^{-4} cm/s or less).
(e) Highly variable.

Determining Cleanup Levels in Bioremediation: Quantitative Structure Activity Relationship Techniques

Vethanayagam Regno J. Arulgnanendran
and Nagamany Nirmalakhandan

ABSTRACT

An important feature in the process of planning and initiating biereme-diation is the quantification of the toxicity of either an individual chem-ical or a group of chemicals when multiple chemicals are involved. A laboratory protocol was developed to test the toxicity of single chemi-cals and mixtures of organic chemicals in a soil medium. Portions of these chemicals are used as a training set to develop Quantitative Structure Activity Relationship (QSAR) models. These predictive mod-els are tested using the chemicals in the testing set, i.e., the remaining chemicals. Moreover mixtures with 10 contaminants in each mixture are tested experimentally to determine joint toxicity for mixtures of chemicals. Using the concepts of Toxic Units, Additivity Index, and Mixture Toxicity Index, the laboratory results are tested for additive, synergistic, or antagonistic effects of the contaminants. These concepts are further validated on mixtures containing eight chemicals that are tested in the laboratory. In addition to the use of the predictive models in evaluating cleanup levels for hazardous waste locations, they are useful to predict microbial toxicity in soils of new chemicals from a con-generic group acting by the same mode of toxicity. These models are applicable when the contaminants act singly or jointly in a mixture.

INTRODUCTION

The current emergence of different bioremediation technologies to the cleanup of contaminated sites has aroused attention in determining the effects of chemical contaminants on the soil medium. Although preliminary data on potential toxicity may be obtained from the available literature, it is imperative that direct toxicity testing be done to assess the problem at hand prior to and subsequent to remediation. The determination of toxicity is one of the essential features in the evaluation of possible remedial action. The toxicity, together with other site characteristics, will determine the type and level of treatment required.

Different bioassays have been developed to assess the toxicity of organic chemicals in the aqueous medium for various test organisms. However, these test procedures may not be used directly to assess the toxicity of a chemical in the soil medium. Under these circumstances a direct approach designed to test the toxicity of these chemicals in the soil would be more acceptable as a test procedure. Because the toxicants are released into the soil medium from time to time, a predictive model to evaluate the microbial toxicity in soils would be a useful tool. These models can be used to flag new chemicals introduced by the various industries for their toxicity, as well as existing chemicals, without extensive laboratory testing.

Predicting Chemical Toxicity: Application of QSAR Techniques

QSAR techniques have been used by the pharmaceutical and pesticide industries in the development of new chemicals. In recent years QSAR techniques have been applied for the prediction of toxicity. The Office of Toxic Substances (OTS) of the U.S. Environmental Protection Agency has used QSAR techniques for hazard assessment since 1981 (Clements et al. 1993). QSAR is based on the premise that a definite relationship exists between the chemical/biological activity and molecular properties of the organic chemicals. Different molecular descriptors have been used by many researchers to derive suitable QSAR models. These molecular descriptors provide quantitative information as to how the modification of a chemical structure results in changes in chemical or biological activity.

By using a set of experimental data as a "training set," QSAR models can be developed correlating the toxicity and the molecular descriptors. Using these QSAR models and the molecular descriptors, toxicity of new chemicals in a "testing set" can be predicted to validate these predictive QSAR models. By employing suitable descriptors of the molecule, and experimentally measured toxicity values, QSAR techniques have been used to predict the toxicity of chemicals in the aqueous medium (Blum and Speece 1990, 1991; Tang et al. 1992; Nirmalakhandan et al. 1994, 1995). In this manner QSAR techniques can supplement and expand the applicability of experimental results to untested chemicals.

EXPERIMENTAL METHODOLOGY

Sandy loam soil was collected from a depth of 15 cm at an agricultural field in Mesilla, New Mexico. The soil was sieved using a 2-mm sieve to remove leaves and other organic material. The measured organic content of the soil was 0.7%. The soil was autoclaved for 7 h daily for 4 d and oven-dried for 3 h at 105°C to sterilize the soil (Tu 1978). A commercially available surrogate culture of microorganisms, Polytox™, was evaluated in the test procedure. An 8-g vial of freeze-dried Polytox™ was dissolved in 280 mL of buffered solution and nutrients prepared according to standard methods. This mixture was supplied with oxygen for 4 h while being stirred continuously. At the end of 4 h, 20 mL of the supernatant from the microbial culture was mixed with 200 g of the autoclaved soil in each of the 600-mL respirometer reactor bottles. To maintain 50% moisture holding capacity, the required amount of water was added to the soil.

Different concentrations of the toxicant, dissolved in 0.5 mL of acetone, were added to each of the reactor bottles, except for the control reactor which received only 0.5 mL of acetone. After mixing the chemicals with the soil, potassium hydroxide pellets were placed in the holder provided in the caps of the reactors. A 12-reactor computer-interfaced respirometer (N—CON Corporation, New York) was used for the assays. These reactors were placed in the respirometer bath maintained at 25°C. The oxygen uptake of each reactor was monitored by the data acquisition system in the respirometer for the next 8 to 10 h.

The concentration of the toxicant causing inhibition of the organisms' respiration by 50%, i.e., IC_{50}, was calculated by comparing the oxygen uptake of each reactor with that of the control that was free of the contaminant. The inhibition percentage at different concentrations of the toxicants was calculated based on the reduction in oxygen uptake rate in each of the reactors with the toxicant in comparison to the toxicant-free control (Elanbaraway et al. 1988).

Joint Toxicity of Mixtures of Chemicals

Equitoxic ratios of the different single chemicals assayed were used to experimentally determine the joint toxicity of 8-component and 10-component mixtures. These mixture combinations were selected at random from the single chemical list of 35 chemicals. These combinations of chemicals at differing concentrations were dissolved in 0.5 mL of acetone and added to the respirometer reactors. The rates of oxygen uptake from these reactors were compared against rates in a control reactor that received 0.5 mL of acetone.

RESULTS AND DISCUSSION

Single-Chemical Experimental Results

Test results from single runs for 35 chemicals by the above experimental technique are given in Table 1. The high r^2 (r = correlation coefficient) values for the dose-response plots listed in Table 1 explain the clear linear variation between chemical concentration and the percentage inhibition of the rate of oxygen uptake for the ranges of values tested. To demonstrate the reproducibility

TABLE 1. Experimental IC$_{50}$ values.

ID #	Chemical Name	Type[a]	IC$_{50}$ [mg/g]	r^2
1	Benzene	ARO	0.51	0.916
2	Toluene	ARO	0.37	0.905
3	*o*-Xylene	ARO	0.22	0.808
4	Ethylbenzene	ARO	0.21	0.921
5	Chlorobenzene	ARO	0.33	0.909
6	1,2-Dichlorobenzene	ARO	0.12	0.819
7	1,3-Dichlorobenzene	ARO	0.14	0.917
8	1,2,4-Trichlorobenzene	ARO	0.24	0.983
9	2,4-Dimethyl phenol	ARO	0.13	0.956
10	Dichloromethane	HAL	0.94	0.722
11	Dibromomethane	HAL	0.68	0.919
12	Carbon tetrachloride	HAL	0.45	0.979
13	1,2-Dichloroethane	HAL	0.51	0.909
14	1,1,1-Trichloroethane	HAL	0.59	0.981
15	1,1,2,2-Tetrachloroethane	HAL	0.12	0.856
16	1,2-Dichloropropane	HAL	0.32	0.987
17	Bromochloromethane	HAL	0.91	0.953
18	Bromodichloromethane	HAL	0.21	0.984
19	Chlorodibromomethane	HAL	0.17	0.849
20	Ethylene dibromide	HAL	0.35	0.962
21	*cis*-1,2-Dichloroethylene	HAL	0.45	0.915
22	Trichloroethylene	HAL	0.56	0.955
23	Tetrachloroethylene	HAL	0.34	0.913
24	Ethanol	AKE	2.59	0.729
25	Propanol	AKE	1.13	0.960
26	Pentanol	AKE	0.45	0.886
27	Octanol	AKE	0.12	0.960
28	*n*-Butyl acetate	AKE	0.45	0.635
29	Isobutyl acetate	AKE	0.57	0.972
30	*n*-Amyl acetate	AKE	0.34	0.945
31	Ethyl acetate	AKE	0.97	0.934
32	Acetone	AKE	4.48	0.975
33	Methyl isobutyl ketone	AKE	0.56	0.828
34	Methyl *n*-propyl ketone	AKE	0.39	0.787
35	Cyclohexanone	AKE	0.95	0.970

(a) ARO—Aromatic; HAL—Halogenated aliphatic; AKE—Alcohols, ketones, and esters.

of the proposed test protocol, duplicate tests were done on 12 of the 35 chemicals. These results yielded an average standard deviation of 0.034 and a coefficient of variation of 0.08 for the 12 chemicals. Moreover, a series of tests were done to evaluate the effect of the soil moisture content. These test results gave an average standard deviation of 0.20 and a coefficient of variation of 0.27.

Single-Chemical QSAR Models

The chemical dose is administered to the soil medium in the form of liquid. From the bulk liquid, the chemical partitions between the soil, the water, the microbial cells, and the headspace in the reactor. In this research it is modeled that the toxic effect on the microorganisms is caused by the concentration of the chemical available as the dissolved form in the soil water. This concentration is determined by mechanistic modeling of the experimental system. Experimental results of 23 test chemicals were used as training set to develop QSAR models using Molecular Connectivity Indices (MCIs) (Kier and Hall 1986). The models are developed for the three congeneric groups of chemicals.

Aromatics:

$$\log IC_{50} (Dissolved) = 0.559 - 1.089\ ^1\chi \tag{1}$$
$$n = 6;\ r = 0.994;\ r^2 = 0.989;\ SE = 0.058.$$

Halogenated Aliphatics:

$$\log IC_{50} (Dissolved) = 0.243 - 1.046\ ^1\chi \tag{2}$$
$$n = 9;\ r = 0.938;\ r^2 = 0.881;\ SE = 0.143$$

Alcohols, Ketones, and Esters:

$$\log IC_{50} (Dissolved) = 0.659 - 1.110\ ^1\chi^V \tag{3}$$
$$n = 8;\ r = 0.997;\ r^2 = 0.994;\ SE = 0.093,$$

where IC_{50} *(Dissolved)* is the concentration of the chemical in the dissolved form that causes 50% inhibition, and $^1\chi\ ^1\chi^V$ are the first-order and first-order-valence MCIs, respectively. The comparison between the experimental and calculated values of the inhibition concentrations is shown in Figure 1.

Prediction of IC_{50} Values for the Testing Set

Twelve chemicals representing three congeneric groups and assayed for toxicity were used as the testing set to validate the QSAR models developed for the 23 chemicals of the training set. Using the model equations by the MCI approach (Equations 1 to 3), the IC_{50} for each of these 12 chemicals was predicted. The comparison of these predicted values and experimental values is shown in Figure 2.

Joint Toxicity of Chemical Mixtures

Results of the mixture toxicity tests are shown in Table 2. For simple additivity, the values of Toxic Units (ΣTU), Additivity Index (AI), and Mixture Toxicity Index (MTI) should be equal to 1, 0, and 1, respectively; Table 2 gives

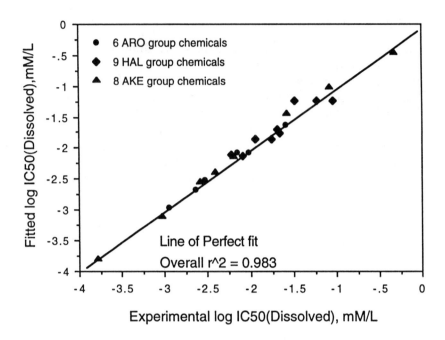

FIGURE 1. Comparison of experimental and QSAR-fitted IC$_{50}$ values.

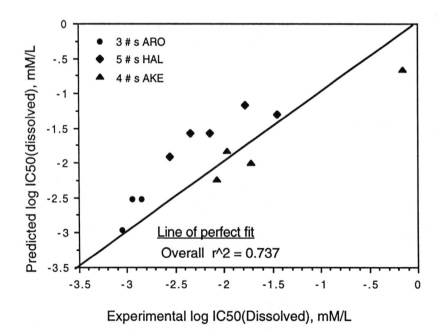

FIGURE 2. Comparison of experimental and QSAR-predicted IC$_{50}$ values for the testing set.

TABLE 2. Toxicity results for the 10- and 8-component mixtures.

Mixture No.	Chemicals in Mixture ID# of Chemical	r^2	$\Sigma TU^{(a)}$	$AI^{(b)}$	$MTI^{(c)}$
	10-Component Mixtures				
10C-1	1,14,9,18,20,17,22,35,33,27	0.924	0.82	0.22	1.09
10C-2	1,10,9,18,20,16,22,35,33,27	0.820	0.85	0.18	1.07
10C-3	5,15,12,18,20,13,22,30,29,27	0.936	0.86	0.16	1.07
10C-4	6,2,11,28,20,13,22,30,29,27	0.991	0.95	0.05	1.02
10C-5	6,2,11,18,29,13,22,35,33,27	0.954	0.99	0.01	1.00
10C-6	7,2,11,18,30,13,22,35,33,27	0.956	1.05	−0.05	0.98
10C-7	1,14,9,18,2,17,22,35,33,27	0.890	1.00	0.00	1.00
10C-8	1,10,9,18,2,16,22,35,33,27	0.887	0.98	0.02	1.01
10C-9	5,3,11,18,2,13,22,34,33,1	0.900	0.93	0.08	1.03
10C-10	5,1,12,2,20,13,22,30,29,27	0.895	0.82	0.22	1.09
10C-11	6,2,11,28,1,13,22,30,29,27	0.879	1.00	0.00	1.00
10C-12	6,2,11,1,29,13,22,35,33,27	0.969	0.98	0.02	1.01
10C-13	7,2,11,1,30,13,22,35,33,27	0.945	0.95	0.05	1.02
10C-14	7,2,11,1,20,31,22,35,33,27	0.955	0.98	0.02	1.01
10C-15	8,2,11,1,22,13,35,5,33,29	0.971	0.90	0.11	1.05
10C-16	8,2,11,22,1,13,21,35,33,29	0.957	1.06	−0.06	0.98
	8-Component Mixtures				
8C-1	9,18,20,17,22,35,33,27	0.986	0.81	0.24	1.10
8C-2	9,18,20,16,22,35,33,27	0.974	1.26	−0.26	0.89
8C-3	5,18,20,13,22,30,29,27	0.923	1.02	−0.02	0.99
8C-4	6,2,11,28,22,30,29,27	0.992	0.83	0.21	1.09
8C-5	6,2,11,18,13,22,33,27	0.943	0.83	0.21	1.09
8C-6	7,2,11,18,30,13,22,35	0.922	0.99	0.01	1.00
8C-7	7,2,11,18,20,22,33,27	0.927	0.93	0.08	1.04
8C-8	8,2,11,18,19,13,21,4	0.977	1.00	0.00	1.00
8C-9	8,2,11,23,21,35,33,26	0.884	1.06	−0.06	0.97
8C-10	1,14,18,2,17,35,33,27	0.921	1.00	0.00	1.00
8C-11	1,10,9,2,16,35,33,27	0.958	1.11	−0.11	0.95
8C-12	5,11,2,13,22,34,33,1	0.942	0.91	0.10	1.05
8C-13	5,1,12,2,13,22,30,29	0.865	0.95	0.05	1.03
8C-14	6,2,11,1,13,22,29,27	1.000	0.96	0.04	1.02
8C-15	2,11,1,29,22,35,33,27	0.992	1.17	−0.17	0.92
8C-16	2,11,1,30,13,22,33,27	0.956	1.06	−0.06	0.97
8C-17	7,2,11,1,31,22,35,27	0.963	1.13	−0.13	0.94
8C-18	8,2,11,1,22,13,35,5	0.996	0.96	0.04	1.02
8C-19	8,2,11,22,1,13,21,35	0.978	0.94	0.06	1.03
	Mean		0.97	0.04	1.02
	$SD^{(d)}$		0.10	0.11	0.05
	$CV^{(e)}$		0.11	2.75	0.05

(a) ΣTU = Summation of Toxicity Units.
(b) AI = Additivity Index.
(c) MTI = Mixture Toxicity Index.
(d) SD = Standard Deviation.
(e) CV = Coefficient of Variation.

average values of $\Sigma TU = 0.97 \pm 0.10$, $AI = 0.04 \pm 0.11$, and $MTI = 1.02 \pm 0.05$. Based on these results it can be concluded that the chemicals exhibit simple additivity when acting jointly in a uniform mixture. In an N-component mixture, as equitoxic ratios of the chemicals are used in the assays, each chemical will exert a toxic effect of 1/N under simple additivity. On the basis of this, the predictions of the Nth chemical in each mixture can be made using the QSAR model equations (Equations 1 to 3), and are shown in Table 3. The comparison of the experimental test results and the QSAR model predictions, based on perfect simple additivity of joint effects of mixtures, is shown in Figure 3.

CONCLUSIONS

The Polytox[TM] surrogate organisms utilized in this study are convenient to use, and microbial toxicity in soil medium can be measured within 8 to 10 h. In almost all chemicals, the variation of the inhibition percentage with the contaminant concentration is explained by the high r^2 values, as shown in

TABLE 3. Prediction of mixture toxicity.

Mixture	Chemicals in Mixture	N^{th} Chemical	Observed	Obs. IC_{50} of Nth Chemical	Observed Concn. of Nth Chemical	Predicted Concn. of Nth Chemical
No.	ID# of Chemical	ID#	ΣTU	mg/L	mg/L	mg/L
	10-Component Mixtures					
10C-1	1,14,18,20,17,22,35,33,27	9	0.82	0.11	0.01	0.01
10C-2	1,10,9,20,16,22,35,33,27	18	0.85	1.19	0.10	0.44
10C-3	5,15,18,20,13,22,30,29,27	12	0.86	1.79	0.15	0.22
10C-4	6,2,11,28,13,22,30,29,27	20	0.95	3.13	0.30	1.27
10C-5	6,2,11,18,13,22,35,33,27	29	0.99	0.45	0.04	0.03
10C-6	2,11,18,30,13,22,35,33,27	7	1.05	0.21	0.02	0.05
10C-7	1,14,9,18,2,22,35,33,27	17	1.00	7.69	0.77	0.76
10C-8	1,10,9,18,2,22,35,33,27	16	0.98	0.91	0.09	0.08
10C-9	5,11,18,2,13,22,34,33,1	3	0.93	0.31	0.03	0.03
10C-10	5,1,12,20,13,22,30,29,27	2	0.82	0.88	0.07	0.08
10C-11	6,2,28,1,13,22,30,29,27	11	1.00	5.63	0.57	1.02
10C-12	6,2,11,1,29,22,35,33,27	13	0.98	2.17	0.21	0.17
10C-13	7,2,11,1,13,22,35,33,27	30	0.95	0.12	0.01	0.01
10C-14	7,2,11,1,20,31,22,35,33,	27	0.98	0.02	0.002	0.002
10C-15	8,2,11,1,22,13,35,5,29	33	0.90	0.85	0.08	0.06
10C-16	8,2,11,22,1,13,21,33,29	35	1.06	1.87	0.20	0.09
	8-Component Mixtures					
8C-1	9,18,20,17,22,35,33	27	0.81	0.02	0.002	0.003
8C-2	9,18,20,16,22,35,27	33	1.26	0.85	0.13	0.07
8C-3	5,18,20,13,22,29,27	30	1.02	0.12	0.01	0.01
8C-4	6,2,11,28,22,30,27	29	0.83	0.45	0.05	0.06
8C-5	6,2,11,18,13,33,27	22	0.83	2.68	0.28	0.33
8C-6	2,11,18,30,13,22,35	7	0.99	0.21	0.03	0.06
8C-7	7,11,18,20,22,33,27	2	0.93	0.88	0.10	0.10
8C-8	8,2,18,19,13,21,4	11	1.00	5.63	0.70	1.27
8C-9	8,2,11,21,35,33,26	23	1.06	0.98	0.13	0.16
8C-10	1,14,2,17,35,33,27	18	1.00	1.19	0.15	0.56
8C-11	1,9,2,16,35,33,27	10	1.11	7.75	1.07	0.62
8C-12	5,11,2,13,22,33,1	34	0.91	0.92	0.10	0.15
8C-13	5,1,12,2,22,30,29	13	0.95	2.17	0.26	0.22
8C-14	2,11,1,13,22,29,27	6	0.96	0.16	0.02	0.06
8C-15	2,11,29,22,35,33,27	1	1.17	2.05	0.30	0.23
8C-16	11,1,30,13,22,33,27	2	1.06	0.88	0.12	0.10
8C-17	7,2,11,1,31,22,27	35	1.13	1.87	0.26	0.12
8C-18	8,2,11,1,22,13,35,	5	0.96	0.78	0.09	0.12
8C-19	8,2,11,22,1,13,35	21	0.94	3.44	0.40	0.60

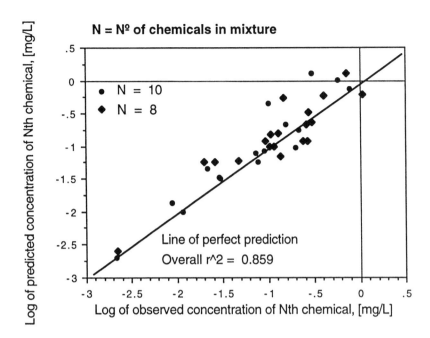

**FIGURE 3. Comparison of observed and predicted Nth chemical concentra-
tions in mixtures.**

Table 1. The correlation between the QSAR-calculated values using the MCI
model equations and experimental results has an overall r^2 of 0.983 for the 23
chemicals in the testing set, indicating the applicability of the models proposed
in this study. The results of the different mixtures indicate a simple additivity
mechanism for the 35 different mixtures assayed. The prediction made by uti-
lizing the MCI models for a chemical selected at random from these mixture
combinations had an overall r^2 of 0.859, indicating the strength of the predictive models for the chemicals involved.

ACKNOWLEDGMENT

The above study was carried out under grant AFOSR-91-0394 from the U.S.
Air Force Office of Scientific Research. Dr. W. Berry, Director, and Dr. W.
Kozumbo, Program Manager.

REFERENCES

Blum, D. J. W., and R. E. Speece. 1990. "Determining Chemical Toxicity to Aquatic Species."
Environmental Science & Technology 24: 284.

Blum, D. J. W. and R. E. Speece. 1991. "Quantitative Structure-Activity Relationships for Chemical Toxicity to Environmental Bacteria." *Ecotoxicology and Environmental Safety* 22: 198.

Clements, R. G., J. V. Nabholz, D. W. Johnson, and M. Zeeman. 1993. "The Use and Application of QSARs in the Office of Toxic Substances for Ecological Hazard Assessment of New Chemicals." In W. G. Landis, J. S. Hughes, and M. A. Lewis (Eds.), *Environmental Toxicology and Risk Assessment, ASTM STP 1179,* American Society for Testing and Materials, Philadelphia, PA, 1179, p. 56.

Elanbaraway, M. T., R. R. Robideau, and S. A. Beach. 1988. "Comparison of Three Rapid Toxicity Test Procedures: Microtox®, Polytox®, and Activated Sludge Respiration Inhibition." *Toxicity Assessment.* Vol. 3, p. 36. John Wiley & Sons Inc., New York, NY.

Kier, L. B., and L. H. Hall. 1986. *Molecular Connectivity in Structure-Activity Analysis.* John Wiley & Sons Inc., New York, NY.

Nirmalakhandan, N., V. R. J. Arulgnanendran, M. Mohshin, S. Bangxin, and F. Cadena,. 1994. "Toxicity of Mixtures of Chemicals to Microorganisms." *Water Research* 28: 543.

Nirmalakhandan, N., B. Sun, V. R. J. Arulgnanendran, M. Mohshin, X. H. Wang, J. Prakash, and N. Hall. 1995. "Analyzing and Modeling Toxicity of Mixtures of Organic Chemicals to Microorganisms." To be published in *Water Science & Technology.*

Tang, N. H., D. J. W. Blum, N. Nirmalakhandan, and R. E. Speece. 1992. "QSAR Parameters for Toxicity of Organic Chemicals to Nitrobactor." *ASCE, Journal of Environmental Engineering 118*: 17.

Tu, C. M., 1978. "Effects of Insecticides on Populations of Microflora, Nitrification and Respiration in Soil." *Commn. in Soil Science and Plant Analysis, 9*(7): 629.

An Application of Adaption-Innovation Theory to Bioremediation

Louise J. Guerin and Turlough F. Guerin

ABSTRACT

This paper provides a discussion of the potential application of the Kirton Adaption-Innovation Inventory (KAI) for assessing the adaptive-innovative cognitive style of individuals and organizations within the bioremediation industry. Human-resource and line managers, or other individuals responsible for staff evaluation, selection, and project planning, should consider using the KAI to assist them in selecting individuals for specific roles requiring either an innovative or adaptive style. The KAI, a measure for assessing adaption-innovation at the individual employee level, is introduced and its potential value in the bioremediation industry is discussed.

INTRODUCTION

Innovation is clearly important in the waste and environmental management industry and specifically in bioremediation since problems in this field often demand novel approaches for their solution (van der Kooi and Guerin 1994). From the vast number of site remediations currently being conducted within the European Union (European Union 1994) and the United States (Kovalick 1994) that employ conventional treatment/management options (e.g., landfilling, vapor stripping), it is evident that there is a need for defining, assessing, optimizing, and managing innovative thinking (and resulting activities) of organizations involved in bioremediation to maximize the rate at which technological developments emerge and are implemented on sites.

The purpose of this paper is to provide a discussion of the usefulness of a measure of adaptive-innovative cognitive styles of individuals and in turn finding the overall adaptiveness-innovativeness of organizations in various aspects of bioremediation. This measure has been used in other fields in the corporate sector and similarly may provide managers in the area of bioremediation with a useful tool for recruiting, selecting particular individuals for specific projects, and effective team building.

DEFINING "INNOVATION"

Typically the term "innovation" has been synonymous with "new," "novel," and "creative" as opposed to not new or not creative (Kirton 1994). However, in this paper we present a theory and measure which is based upon the premise that all individuals are creative and can therefore generate novel solutions. However, an individual generates novel solutions in one of two ways and these processes are either "adaptive" or "innovative." An innovative solution has an element of "breakthrough" about it and involves a discontinuous cognitive approach. An innovative solution would be generated by working to the edges of a paradigm or even breaking out of the paradigm's boundary. An example of an innovative approach is changing from landfilling of wastes to biological treatment (bioremediation). Conversely, an example of an adaption in the bioremediation industry would most likely be an application of an existing technology or knowledge such as the use of biostimulation instead of bioaugmentation or the application of composting to the treatment of hazardous organic compounds in a contaminated soil.

TECHNOLOGICAL DEVELOPMENTS IN THE BIOREMEDIATION INDUSTRY

In recent years, considerable progress has been made in the field of biological treatment of hazardous wastes and contaminated soil. Many of the successful international field-scale studies have been documented in Hinchee et al. (1994), and those conducted in Australia by Guerin et al. (1994) and Rhodes et al. (1994).

Current technological developments in the bioremediation industry include the development and trials of novel composting materials and organic amendments for enhancing the degradation of recalcitrant organics, the use of controlled release nutrients, the application of mixed cultures of microorganisms in bioaugmentation processes, a general trend to use biostimulation approaches wherever possible, improved oxygen delivery systems, and other advancements in engineering aspects for both in situ and ex situ treatment facilities. The use of surfactants as a possible means for increasing the bioavailability of soil- and sediment-bound pollutants and the application of plants and plant-microbial systems to soil bioremediation are examples of others. Some of these are listed in Table 1.

Many technologies in the bioremediation industry are often so new that their usefulness (and validity) has not been fully determined. Given the fast rate of emergence of these new technologies and approaches, the United States Environmental Protection Agency (EPA) has developed a means of assessing "innovative" technologies. Specifically, in the United States this is being achieved through the SITE (or Superfund Innovative Technology Evaluation Program). There is presently no equivalent organization in Australia.

TABLE 1. Recent technological developments in bioremediation of contaminated soil.[a]

Technological Development	Status (Implemented or Developmental)	Innovation or Adaption (using A-I Theory)	Major Advantage/Constraint
Clear demonstration of biological contribution in bioremediation processes	Implemented	Adaption	Increasing confidence in bioremediation
Real-time monitoring of bioremediation process data	Implemented	Adaption	Rapid response to changing process conditions
Ammonia injection	Implemented	Adaption	Accelerates removal of low concentrations of hydrocarbons
Controlled-release nutrients	Implemented	Adaption	Particularly effective for ex situ bioremediation of TPH- and VOC-contaminated soil
Application of composting to treatment of contaminated soil	Implemented	Adaption	Treatment of wide range of organics and fast rates of degradation in cool environments; potential for residual phytotoxicity
Surfactants	Developmental	Adaption	Limited evidence for efficacy in either feasibility or field-scale trials
Hydrogen peroxide	Developmental	Adaption	Potential to increase groundwater oxygen concentrations, but field data are limited
Application of materials handling equipment to bioremediation engineering	Developmental	Adaption	Concentrating contaminants into selected soil fractions; potential synergy identified with the mining industry
Concept of using metal hyperaccumulating plants	Developmental	Innovation	Limited evidence from feasibility trials that concept will be applicable to anthropogenic sources of metal contamination
Bioventing	Implemented	Innovation	Maximizes biodegradation and minimizes VOC losses due to volatilization

(a) Source: Some of the information on implemented technologies has been compiled from field-scale bioremediation trials conducted by Minenco Pty Ltd. Bioremediation Services (one of the CRA group of companies) and from Guerin (1994), Guerin et al. (1994), Kelly and Guerin (1994), Rhodes et al. (1994), Hinchee et al. (1994), and Means and Hinchee (1994).

Although considerable technological advances have been made in bioremediation, there are particular aspects of the industry where innovation is required if further gains are to be made. These specifically include increasing the bioavailability of tightly bound and potentially biodegradable contaminants, improved indicators for predicting ability to scale-up from feasibility trials, and the development of new processes capable of increasing the cometabolic degradation of highly recalcitrant organics. Some of these issues have been raised by Alexander (1991). The treatment of metals poses a considerable challenge for bioremediation (Means and Hinchee 1994) and is reflected in the U.S. EPA's Superfund Program, where the preferred treatment option is primarily solidification/stabilization, even though the more novel technologies (although not biological) of soil washing and in situ flushing are being employed to a lesser extent (Kovalick 1994).

MEASURING THE INNOVATIVE-ADAPTIVE COGNITIVE STYLE OF INDIVIDUALS

To maximize organizational innovation or organizational adaption where it is relevant, it would be beneficial for the industry to be able to measure this aspect of individual cognitive style in the workplace. This information could then be used by research planners, and human-resource and project or line managers to optimize personnel effectiveness and placement. The measure currently being used for assessing individual adaption-innovation in other industries is the KAI. This measure assesses individual adaptive-innovative cognitive styles within the workplace; that is, an individual's preferred style of problem solving and decision making.

Adaption-Innovation (A-I) Theory

At the basis of adaption-innovation theory exists a dimension "adaption" and "innovation" where the two are extremes of a bipolar dimension. Individuals can be located at any point along this dimension and this adaptive-innovative individual style has been shown to be normally distributed in the population. The dimension of adaption-innovation reflects an individual's cognitive style underlying creativity, problem-solving and decision-making behavior (Kirton 1976, 1994). Specifically, adaption-innovation theory explains this behavior within an organizational setting.

Cognitive styles are personality traits, where these traits are variables or dimensions along which individuals differ. Cognitive style has been defined as "consistent individual differences in preferred ways of organizing information" (Kirton and De Ciantis 1986). It refers to the manner or mode of cognition, that is, the way in which a person thinks, approaches problems, or adopts strategies to solve problems. It has been found to be independent of levels of ability, skills, intelligence, and levels of cognitive complexity. It has also been found to be stable over time and situations, and, as a consequence, remains largely unresponsive to specific training.

The Kirton Adaption-Innovation Inventory (KAI)

The KAI is the actual measure of the dimension adaption-innovation and was designed specifically for use in the work environment. It establishes at which point along the continuum an individual can be located with respect to cognitive style.

The KAI was developed in the United Kingdom in the 1970s and is a measure with international normative data and a vast amount of research supporting its validity and reliability. It is a "pencil-and-paper" instrument of 32 items and is not a timed test but on average takes 10 min to complete. The inventory has to be administered and scored by an accredited user and scores range from 32 to 160, where these extreme scores represent the extreme styles, adaptor and innovator. Since the dimension of adaption-innovation is normally distributed it would be extremely rare for someone to be either a pure innovator or adaptor.

Use of KAI. The KAI is being used internationally in a variety of industries in assessment batteries for personal and professional development, in consulting with employees on creativity and problem solving and its relevance to organizations, to complement Total Quality Management (TQM) programs, in managing change, for continuous improvement, and for recruitment and effective structuring and use of work teams.

Administering the KAI. To administer the KAI the administrator should allow 3 h for an introductory information session for those undertaking the KAI. Following this the administrator should allow at least 30 min per participant to score each test and to provide one-to-one feedback for the individual participant. The average cost of the inventory itself is less than Australian $10 per person. The cost to management in terms of putting individuals through this program would be a 3-h information session for all those participating, approximately 10 min for each individual to complete the inventory, and approximately 20 min for each individual to receive feedback.

Characteristics of Adaptors and Innovators

In adaption-innovation theory, the terms "adaptors" and "innovators" are used to denote hypothetical types at each end of the continuum. However, in reality, individuals are located along the entire continuum. When confronted with a problem, adaptors turn to conventional procedures and the consensus of the group, whereas innovators will attempt to restructure the problem by approaching it from a new angle. Innovators prefer to generate new and different solutions to problems, which disrupts rather than preserves the familiar patterns in which problems exist. That is, innovators are characteristically less concerned with the maintenance of paradigms, and so their creativity is more likely to lead to a new paradigm or a paradigm-switch. They also tend to incorporate the context of a problem into the problem itself and seek to change the paradigm as part of their solutions.

In contrast, adaptors prefer familiar methods and use existing principles to solve problems. Adaptors are distinguished from innovators in that they seek to solve problems in a way that supports, refines, and extends generally accepted paradigms. They tend to cope with novel stimuli by assuming that a relevant paradigm has the power to resolve that problem. Although adaptors and innovators are at variance in their cognitive styles, that is, in their problem-solving and decision-making behavior, they may be equally creative, and both creative styles may be equally appropriate and effective, depending on the situation.

The adaptor is characterized by precision, reliability, and efficiency, and is prudent, disciplined, and conforming. She or he seeks to solve problems in tried and understood ways, rarely challenges rules, is sensitive to other people, and is able to maintain group cohesion and cooperation when working with others. Adaptors seem to prefer the production of (as distinct from being capable of producing) fewer original ideas than innovators. In contrast, the innovator is characterized as being undisciplined, thinking tangentially, and, in problem solving, discovering many avenues of solutions. She or he often challenges rules, possesses the ability to bring about radical change and has little respect for past custom. When working with others, the extreme innovator is insensitive to others and often threatens group cohesion. Other personality characteristics associated with innovators include being impatient with detail, less concerned with being systematic and methodical, and more easily bored with routine.

The Value of Adaptors and Innovators

The behavior patterns of adaptors and innovators have far-reaching implications for the workplace, which is why adaption-innovation theory has played an expanding role in the explanation of organizational behavior. According to adaption-innovation theory, individuals bring entirely different viewpoints and solutions to administrative and organizational problems. The task of getting solutions accepted can be considerably greater for innovators than for adaptors, because innovative behavior is by definition more radical and therefore not likely to meet with approval when first introduced. However, in problem-solving, adaptors exhibit greater constraint, regard for the opinions of others, and other attributes of "immediate value to systems and bureaucracies" (Kirton 1978).

The differences in cognitive styles (innovation-adaption) not only affect how individuals prefer to work but also how they perceive those who have the contrasting style. There are dangers that adaptors and innovators may not only disagree on appropriate projected action but hold negative views of each other. They also have a tendency not to see each others' points of view. Both are needed in organizations if only to cover each others weaknesses but, of the two, the adaptor has a more privileged position because it is the adaptive mode that must prevail more consistently.

Therefore, an advantage of explaining this theory in the workplace is that better understanding and mutual respect will bring about improved collaboration and communication between these individuals. Being able to recognize one's own cognitive style and those of one's colleagues is important in resolving situations which require problem-solving and decision-making.

INNOVATIVE AND ADAPTIVE ORGANIZATIONS

Organizations become adaptive or innovative as a result of the personnel in that organization. Although organizations exert demands on their personnel, it is unlikely that any change in behavior due to these demands will be so profound as to alter something as basic as cognitive style. Cognitive style has been shown not to change over time, yet individuals who are expected to exhibit behavior not in accord with their own cognitive style may exhibit "coping behavior" (Kirton 1994; Foxall and Payne 1989).

To fit into an organization or an environment within an organization, individuals can only make superficial changes in their behavior. If individuals find themselves in a cognitive climate that is not suited to their own style, the individual will either find a niche within that organization, make changes to the job to get a better person-environment fit, or, as a final outcome, leave the organization (Hayward and Everett 1983; Kirton and McCarthy 1988). The individual who is "trapped and unhappy" is required to make the most accommodation between his or her own preference and the group consensus (Kirton 1994). Therefore, the issue is both of person-environment fit and maximizing effectiveness of either an adaptor or innovator without requiring them to stifle or cover up their preferred styles of problem solving and making decisions.

The KAI mean scores for certain occupational populations have been seen to aggregate on either side of the general population mean, while still retaining a well-spread distribution (Holland et al. 1991). Accountants, bank employees, programmers, local government authorities, and those in various forms of production and maintenance have been shown to locate toward the adaptive end of the continuum. Whereas those in research and development, planning, personnel, sales, and marketing have been located toward the innovative end of the continuum. Groups that tend to be amorphous in nature are located toward the midpoint of the continuum.

Even departments in organizations display a tendency toward either adaption or innovation depending on the functions they perform. The demands of tasks undertaken by an organization require certain cognitive styles. Therefore, if selection procedures are adequate, the personnel should reflect the organization's ethos of either adaptiveness or innovativeness. It has been found that individuals working in departments primarily concerned with their departments' own internal processes, produced a mean score that was more adaptive than the general population mean. Conversely, the mean scores of groups at the interface between a number of departments or between their departments and outside of the organization were more innovative.

CONCLUSIONS AND RECOMMENDATIONS FOR APPLIED RESEARCH

Organizations involved in research on and development and implementation of bioremediation processes are being forced to be increasingly efficient so that they remain competitive. Therefore, assessing the adaptiveness-innovativeness of individuals and organizations involved in these broad areas of bioremediation could be useful for improving the overall competitiveness of this industry. The current paper has investigated the potential application of the KAI, an inventory currently used internationally for assessing innovation in the workplace, for use in the bioremediation industry. Line and human resource managers should consider applying the KAI for selecting individuals for specific jobs requiring either an innovative or adaptive cognitive style.

The role of the KAI needs to be explored in this industry. For example, it is likely to be an important tool for selecting the most appropriate individuals for specific tasks in various organizations. The KAI is also likely to have long-term implications for organizational change. It should be appreciated that a group of only high innovators (according to the adaption-innovation theory) in a bioremediation research or development team, is unlikely to provide the means for achieving a successful outcome, and will not benefit from the methodical and efficient implementation that adaptors could offer. Rather, experience from previous applications of the KAI throughout other industries suggests that a balance of innovators and adaptors provides the optimum team for maximum productivity. However, in a R&D project team environment there should be a larger number of innovative individuals with some adaptive individuals to drive the implementation of novel technological developments. Attention now needs to be given to defining the appropriate balance of cognitive styles within specialized departments or organizations.

Line and human-resource managers in the bioremediation industry should be aware of the KAI's advantages and potential for assessing the innovativeness of their personnel and therefore of their department or organization.

ACKNOWLEDGMENTS

The senior author (L. J. Guerin) from Innovation Assessment & Research would like to acknowledge Professor Michael Kirton, Director, Occupational Research Centre, United Kingdom for assistance and accreditation using the KAI.

REFERENCES

Alexander, M. 1991. "Research Needs in Bioremediation." *Environmental Science and Technology* 25 (12): 1972-1973.

European Union. 1994. *International Workshop—Contaminated Sites in the European Union: Policies and Strategies,* Federal Ministry for the Environment, Nature Conservation and Nuclear Safety, Bonn, Germany.

Foxall, G.R., and A.F. Payne. 1989. "Adaptors and Innovators in Organizations: A Cross-Cultural Study of the Cognitive Styles of Managerial Functions and Subfunctions." *Human Relations 42*: 639-649.

Guerin, T.F. 1994. "An Appraisal of the Technical Constraints to the Bioremediation of Contaminated Soils-Implications for the Disposal of Hazardous Wastes." Submission to the Senate Standing Committee on Environment, Recreation and the Arts Inquiry into Waste Disposal. p. 20. Department of Environment, Recreation and the Arts, Canberra, Australia.

Guerin, T.F., S.H. Rhodes, B.C. Kelley, and P.C. Peck. 1994. "The Application of Bioremediation to the Clean-up of Contaminated Sites: The Potential and Limitations." *Proceedings of the Symposium on Contaminated Soil,* pp. 12. Royal Australian Chemical Institute and the University of Newcastle, Australia.

Hayward, G., and C. Everett. 1983. "Adaptors and Innovators: Data from the Kirton Adaptor-Innovator Inventory." *Journal of Occupational Psychology 56*: 339-342.

Hinchee, R.E., D.B. Anderson, F. Blaine Metting, and G.D. Sayles (Eds.). 1994. *Applied Biotechnology for Site Remediation.* Lewis Publishers, Boca Raton, FL.

Holland, P. A., Bowskill, I., and A. Bailey. 1991. "Adaptors and Innovators: Selection Versus Induction." *Psychological Reports 68*: 1283-1290.

Kelly, R.J., and T.F. Guerin. 1994. "The Potential of Using Plants as an Innovative Approach to Soil Decontamination." pp. 591-598. *2nd National Hazardous and Solid Waste Convention,* Australian Waste Water Association, Sydney, Australia.

Kirton, M.J. 1976. "Adaptors and Innovators: A Description and Measure." *Journal of Applied Psychology 61*: 622-629.

Kirton, M.J. 1978. "Have Adaptors and Innovators Equal Levels of Creativity." *Psychological Reports 42*: 695-698.

Kirton, M.J. (Ed.). 1994. *Adaptors and Innovators: Styles of Creativity and Problem Solving.* Routledge, London.

Kirton, M.J. and S.M. De Ciantis. 1986. "Cognitive Style and Personality: The Kirton Adaption-Innovation and Cattell's Sixteen Personality Factor Inventories." *Personality and Individual Differences 7*: 141-146.

Kirton, M.J. and R.M. McCarthy. 1988. Cognitive Climate and Organizations. *Journal of Occupational Psychology 61*: 175-184.

Kovalick, W.W. 1994. Perspectives on Innovative Remediation Technologies in the U.S. Superfund Program. *2nd National Hazardous and Solid Waste Convention,* pp. 11-16. Australian Waste Water Association, Sydney, Australia.

Means, J.L., and R.E. Hinchee (Eds.). 1994. *Emerging Technology for Bioremediation of Metals.* Lewis Publishers, Boca Raton, FL.

Rhodes, S.H., B.C. Kelley, P.C. Peck, D. Longford, and T.F. Guerin. 1994. " The Implementation of Bioremediation in the Mining Industry." *Biomine '94,* pp. 13.1-13.11. Australian Mineral Foundation, Adelaide, Australia.

Van der Kooi, L.J., and T.F. Guerin. 1994. Innovation in Waste and Environmental Management: Measuring and Assessing Innovation in the Workplace. *2nd National Hazardous and Solid Waste Convention,* pp. 576-582. Australian Waste Water Association, Sydney, Australia.

A Bench-Scale Biotreatability Methodology to Evaluate Field Bioremediation

Amy G. Saberiyan, James R. MacPherson, Jr., Jeffrey S. Andrilenas, Roy Moore, and Alan J. Pruess

ABSTRACT

A bench-scale biotreatability methodology was designed to assess field bioremediation of contaminated soil samples. This methodology was performed successfully on soil samples from more than 40 sites. The methodology is composed of two phases, characterization and experimentation. The first phase is physical, chemical, and biological characterization of the contaminated soil sample. This phase determines soil parameters, contaminant type, presence of indigenous contaminant-degrading bacteria, and bacterial population size. The second phase, experimentation, consists of a respirometry test to measure the growth of microbes indirectly (via generation of CO_2) and the consumption of their food source directly (via contaminant loss). Based on a Monod kinetic analysis, the half-life of a contaminant can be calculated. Abiotic losses are accounted for based on a control test. The contaminant molecular structure is used to generate a stoichiometric equation. The stoichiometric equation yields a theoretical ratio for mg of contaminant degraded per mg of CO_2 produced. Data collected from the respirometry test are compared to theoretical values to evaluate bioremediation feasibility.

INTRODUCTION

The science of remediation has developed significantly in the past several years. Collection of certain data is critical to defensible design, engineering, and installation of remediation systems. Such data include biodegradation parameters as discussed here.

A focused, consistent testing approach is necessary to develop a reliable database to support selection of cost-effective treatment technologies. In the case of biodegradable contaminants, this approach is accomplished through a biofeasibility and/or biotreatability study. Although biofeasibility and biotreatability are commonly used interchangeably, they have slightly different meanings. Biofeasibility usually applies to a series of tests and subsequent data review to evaluate bioremediation as a remedial option. Biotreatability applies to studies that fit into a database used to consistently design and monitor successful remediation processes.

The combined goals of these studies are to (1) identify contaminant degraders and enumerate their populations, (2) estimate the contaminant half-life and evaluate bioremediation performance, (3) estimate the time required to achieve remedial goals, (4) identify amendments to the system that enhance biological performance, and (5) generate predictive information to support planning the monitoring phase of remediation. The focus here is a consistent methodology to investigate the feasibility of aerobic bioremediation for contaminated soil or water. Due to inherent variability in site conditions, each biotreatability study is tailored to meet the requirements of a specific project.

There is currently debate on whether biotreatability studies are useful for selection of bioremediation as a cost-effective remediation process. As in any other scientific research or engineering study, results obtained from biotreatability studies have their limitations. Although conclusions from these studies are based on experimental results and data correlation, they provide useful information to optimize and monitor the efficiency of a bioremediation process. Biotreatability methodology (as described herein) has been used to investigate bioremediation feasibility for more than 40 sites. Five case studies are selected and data obtained are summarized in Table 1. The biotreatability results have been comparable with actual site operation and performance for all sites reporting data. Following are general trends and typical project circumstances to perform a biotreatability study.

PHASE 1: CHARACTERIZATION

A series of physical, chemical, and biological screening is conducted on soil and/or water samples to characterize relevant parameters. These include contaminant and bacterial identifications, as well as other parameters of interest.

Petroleum products contain a wide array of hydrocarbon chain lengths and bond arrangements. The biodegradation rate of these molecules is dependent on the number of carbons and the chemical C-C bonds. Identification of the contaminant(s) of concern, or a representative carbon chain length for a mixture, is crucial for data interpretation when applying a stoichiometric relationship between contaminant reduction and CO_2 production. Also important is the contaminant concentration.

Bacterial characterization and enumeration are used to evaluate the presence of indigenous petroleum-degrading bacteria and their initial population, or colony-forming units (CFU). Bacterial characterization involves microscopic examination, and oxidase and gram stain tests.

The samples' initial and final total organic carbon (TOC) content, levels of nutrients (comprised of nitrogen, phosphorus, and potassium), pH, moisture level, porosity, and grain size are obtained. Nutrient analysis and amendment recommendations are discussed in more detail later in this paper.

TABLE 1. Selected case studies.

	Case study # 1	Case study # 2	Case study # 3	Case study # 4	Case study # 5
Media	soil	soil	water	soil	soil
Contaminant type	diesel	diesel/motor oil	gasoline	gasoline	diesel
Carbon-chain range	C_{11}-C_{24}	C_{11}-C_{35}	C_4-C_{10}	C_4-C_{10}	C_{11}-C_{24}
Initial TPH[(a)] concentration (mg/kg or L)	52	2,300	1,300	53	3,400
Study duration (day)	28	35	12	18	28
Contaminant reduction (mg/kg or L)	43	1,907	1,100	7	2,603
Total CO_2 production (mg/kg)	46	1,673	NA[(b)]	2,289	3,025
Empirical ratio (TPH/CO_2)	0.93	1.4	NA	0.003	0.76
Theoretical ratio (TPH/CO_2)	1.27	1.27	1.043	1.043	1.27
Contaminant half-life (day)	11	12	5	NA	10
Rate of reduction (-1/day)	-0.065	-0.056	-0.137	NA	-0.07
Bacterial density (CFU/g or L)	4.6×10^7	1.09×10^5	2.24×10^4	1.92×10^7	3.3×10^5
Cleanup goal (mg/kg or L)	10	200	1	10	200
Is bioremediation feasible?	yes	yes	yes	no	yes

(a) TPH = total petroleum hydrocarbon.   (b) NA = not applicable.

PHASE 2: EXPERIMENTAL DESIGN

A combined kinetic analysis and stoichiometric approach to the bioremediation provides valuable insight toward effective engineering design, operation, and monitoring. Growth of bacteria in the presence of a contaminant, although a good indicator, is insufficient to derive a degradation rate. A degradation rate is calculated from changes in contaminant and CO_2 concentrations that are then used in a Monod model. The empirical degradation ratio is compared to the theoretical stoichiometric values (Table 1).

Experimental Setup

At least three experimental cells are used. Each cell contains the same mass of soil or water sample (≥ 1 kg). One subsample is biologically sterilized for use in a control cell(s). The control provides the rate of contaminant removal by abiotic processes such as volatilization or sorption. Another cell(s) is treated with a nutrient-free amendment to generate a baseline rate of biodegradation. Data obtained from the nutrient-free cell(s) are useful for evaluating in situ bioremediation where nutrient, microbe, or cometabolite amendment is infeasible. Finally, a cell(s) is nutrient-amended at levels based on Phase 1 results. To optimize biodegradation, other parameters such as pH, oxygen, and moisture (for soil samples) are adjusted for this cell. Soil or water from each cell is sampled and analyzed for contaminant concentrations over time. CO_2 is captured and measured throughout the study. Conductivity, temperature, cell airflow rate, and O_2 consumption are also monitored. The experiment duration typically is 4 to 6 weeks for soil samples and 1 to 3 weeks for water samples.

Stoichiometric Relationship

A single oxidation-reduction reaction cannot be used for biodegradation analysis. Because some of the carbon becomes cell material and some TOC is degraded, there is a more complex stoichiometric relationship between the contaminant reduction and CO_2 generation. An average carbon chain length is determined from Phase 1 analysis (e.g., C_7H_{12} and $C_{14}H_{24}$ for gasoline and diesel, respectively). The chemical structure of the biomass is assumed to average $C_5H_7O_2N$ per mole. Table 2 contains a stoichiometric relationship for biodegradation of gasoline (McCarty 1988).

Kinetic Model

The Monod equation is commonly used to model substrate degradation and microbial growth. The Monod equation assumes a single substrate and single microorganism (Alexander and Scow 1989). In nature there are multiple substrates and microorganisms. The Monod equation is selected for ease in analyzing data and offers adequate accuracy.

A single chemical formula represents the contaminant(s) based on the average carbon and hydrogen content of the hydrocarbon chains (Riser-Roberts 1992a). The Monod equation is a first-order kinetic reaction:

$$dC/dt = -k_m \times X/K_s \times C_0 \qquad (1)$$

where,

C_0 = initial contaminant concentration (mg/kg)

TABLE 2. Stoichiometic relationship.

Energy reaction
ED * = C_7H_{12} (assumed to be representative of gasoline)
EA = O_2

Electron donor (ED) half-reaction:
C_7H_{12} + 14 H_2O → $7CO_2$ + $40H^+$ + $40e^-$

Electron acceptor (EA) half-reaction:
$4H^+$ + $4\bar{e}$ + O_2 → $2H_2O$

Energy reaction = ED + EA
1/40 C_7H_{12} + 14/40 H_2O →   7/40 CO_2 + H^+ + $\bar{e}$
H^+ + $\bar{e}$ + 1/4 O_2 →   2/4 H_2O

1/40 C_7H_{12} + 7/20 H_2O + 1/4 O_2 →   7/40 CO_2 + 1/2 H_2O

Synthesis reaction: This reaction represents the amount of microbial cells produced in bioremediation processes using gasoline and ammonium.

1/4 CO_2 + 1/20 NH_3 + H^+ $\bar{e}$ →   1/20 $C_5H_7O_2N$ + 2/5 H_2O

1/40 C_7H_{12} + 14/40 H_2O →   7/40 CO_2 + H^+ + $\bar{e}$

1/40 C_7H_{12} + 1/20 NH_3 + 3/40 CO_2 →   1/20 $C_5H_7O_2N$ + 1/20 H_2O

Overall reaction = A (energy reaction) + synthesis reaction:

Assume A = 1

1/40 C_7H_{12} + 7/20 H_2O + 1/4 O_2 →   7/40 CO_2 + 1/2 H_2O

1/40 C_7H_{12} + 1/20 NH_3 + 3/40 CO_2 →   1/20 $C_5H_7O_2N$ + 1/20 H_2O

1/20 $C_7 H_{12}$ + 1/4 O_2 + 1/20 NH3   →   1/10 CO_2 + 1/20 $C_5H_7O_2N$ + 1/5 H_2O
or
C_7H_{12} + 5 O_2 + NH_3 →   2 CO_2 + $C_5H_7O_2N$ + 4 H_2O

For each mole of C_7H_{12}:
   5 mole of O_2 utilized
   1 mole of NH_3 utilized
   2 mole of CO_2 produced
   1 mole of biomass produced

C   = contaminant concentration at time t (mg/kg)
k_m = maximum substrate utilization rate (day^{-1})
K_s = half-velocity coefficient (mg/kg)
X   = microbial concentration (mg/kg)
t   = time (days)

Assuming,

$$K = k_m \times X/K_s \qquad\qquad (2)$$

Then,

$$\ln C/C_0 = -K \times t \qquad\qquad (3)$$

where   K = degradation rate constant.

The value K is measured empirically from the biotreatability results.

Data Interpretation

Nutrient Analysis. The moles of nitrogen required to theoretically convert a mole of contaminant is determined from the stoichiometric relationship (Table 2). The moles of phosphorus and potassium required are estimated assuming a ratio of 10:1:0.1 (Riser-Roberts 1992b) for nitrogen:phosphorus: potassium (N:P:K).

Contaminant Half-Life Analysis. Control-corrected concentrations of the contaminant at time t relative to the initial concentration are plotted over time on a logarithmic scale. Linear regression analysis yields the slope of the line (K). The contaminant biodegradation half-life then can be calculated from the K value ($t^{\frac{1}{2}} = \ln 2/K$).

Contaminant/CO_2 Ratio. A theoretical ratio of contaminant degraded per CO_2 produced can be estimated based on the stoichiometric relationship shown in Table 2 (e.g., for gasoline 1.04, for diesel 1.27). Empirical data are compared with these theoretical ratios (Table 1). Using initial and final TOC contents of the soil or water sample and the theoretical mass ratios, it is possible to estimate how much CO_2 production is due to TOC degradation and what portion is due to contaminant degradation.

CONCLUSIONS

Biotreatability studies have been successfully used to predict bioremediation as an effective treatment option. The methodology described here (1) identifies contaminant-degrading microbes, (2) enumerates their populations,

(3) identifies limiting nutrients or other factors, (4) measures contaminant half-life, and (5) measures a contaminant/CO_2 ratio. Foremost, these provide a good measure of whether biodegradation occurs at a feasible rate at a given site. These data are also used to determine if amendments (e.g., nutrients or non-native microbes) will enhance biological performance. The half-life enables estimates of the time required to achieve remedial goals for cost projections. Half-life and contaminant/CO_2 ratio values allow periodic evaluation of remediation system performance. Additionally, these data indicate what parameters to monitor during remediation.

The information gained from a biotreatability study has been used to make a defensible case to regulators for selecting, or excluding, bioremediation as a cost-effective treatment technology. Although some site investigations generate sufficient data to derive an integrated degradation rate, or intrinsic attenuation rate, these are the exceptions. Given the often limited information available, biotreatability studies provide crucial information to evaluate the feasibility of treatment by bioremediation, a recurrent, cost-effective technology.

REFERENCES

Alexander, M., and K. M. Scow. 1989. "Kinetics of Biodegradation in Soil." *Reactions and Movement of Organic Chemicals in Soils*, pp. 243-264. Soil Science Society of America and American Society of Agronomy.

McCarty, P. L. 1988. "Bioengineering Issues Related to In Situ Remediation of Contaminated Soil and Groundwater." *Environmental Biotechnology*, pp. 128-147. Plenum Publishing Corporation.

Riser-Roberts, E. 1992a. "Organic Compounds in Refined Fuels and Fuel Oils." *Bioremediation of Petroleum Contaminated Sites*, pp. A9-A16. CRC Press, Inc., Boca Raton, FL.

Riser-Roberts, E. 1992b. "Nutrients." *Bioremediation of Petroleum Contaminated Sites*, pp. A9-A164. CRC Press, Inc., Boca Raton, FL.

Monitoring Bioremediation of Weathered Diesel NAPL Using Oxygen Depletion Profiles

Gregory B. Davis, Colin D. Johnston,
Bradley M. Patterson, Christopher Barber,
Marlene Bennett, Alan Sheehy, and Michael Dunbavan

ABSTRACT

Semicontinuous logging of oxygen concentrations at multiple depths has been used to evaluate the progress of an in situ bioremediation trial at a site contaminated by weathered diesel nonaqueous-phase liquid (NAPL). The evaluation trial consisted of periodic addition of nutrients and aeration of a 100-m² trial plot. During the bioremediation trial, aeration was stopped periodically, and decreases in dissolved and gaseous oxygen concentrations were monitored using data loggers attached to in situ oxygen sensors placed at multiple depths above and within a thin NAPL-contaminated zone. Oxygen usage rate coefficients were determined by fitting zero- and first-order rate equations to the oxygen depletion curves. For nutrient-amended sites within the trial plot, estimates of oxygen usage rate coefficients were significantly higher than estimates from unamended sites. These rates also converted to NAPL degradation rates, comparable to those achieved in previous studies, despite the high concentrations and weathered state of the NAPL at this test site.

INTRODUCTION

Monitoring and validating the effectiveness of in situ strategies to bioremediate NAPL contamination of an aquifer is problematic. Periodic recovery of cored aquifer material and analysis to determine NAPL content changes is costly, is time consuming, and does not allow repeat sampling at the same location. Despite this, coring is a direct measure of the effects of a remediation strategy and an essential final validation step.

An often-used, indirect measure of the effectiveness of a bioremediation strategy is to assess oxygen utilization rates (Pijls et al. 1994) or carbon dioxide production rates (Wood et al. 1993), which allow estimation of NAPL degradation rates (Hinchee & Ong 1992). Difficulties with this approach include the need for frequent, repeated sampling of soil gas from boreholes to determine soil gas composition changes with time; the need to monitor at several depths to determine the range of NAPL degradation rates at different depth intervals above and in the NAPL-contaminated zone; and the need to determine air-filled porosities to calculate NAPL degradation rates from rates of decrease of oxygen concentration.

These aspects were investigated as part of a 6-month field trial carried out to test the feasibility of in situ bioremediation of weathered diesel fuel in a shallow sandy aquifer in Perth, Western Australia. Semicontinuous automated logging of oxygen concentrations was carried out at multiple depths. The effect of the location of the water table and the variability of air-filled porosity in the zone of NAPL contamination on NAPL degradation rate estimates also were examined.

SITE DESCRIPTION AND REMEDIATION STRATEGY

The stratigraphy of the shallow, unconfined aquifer at the field site is variable, with thin layers of interbedded fine- and coarse-grained sand in the interval of NAPL contamination, which typically is 3.9 to 4.5 m below ground. Peaks (5 to 26% by weight) in the NAPL distributions were shown by Johnston & Patterson (1994) to be located at or immediately above the water table. The interval of NAPL contamination coincided with the interval of water table fluctuation (about 0.5 m annually). Chromatographic analysis of the NAPL material indicated it to be highly weathered (depleted of N-alkanes), principally consisting of C-10 to C-25 compounds.

The bioremediation trial consisted of nutrient addition and aeration of a 100-m^2 plot to induce increased microbial degradation of the weathered diesel. During aeration, the water table was drawn down to approximately 4.5 m below ground to expose the NAPL-contaminated zone to air. Aeration was halted periodically (often after 140 h of aeration) to measure oxygen concentration decreases over time and estimate oxygen usage rate coefficients. Two locations were chosen for measuring oxygen usage rate coefficients: one in an area of the trial plot that was amended with nutrients (MI1), and the other in a control area where no nutrients were added (MI3).

METHODS

To enable in situ monitoring of dissolved or gaseous oxygen concentrations, electrochemical oxygen sensors (Takeuchi & Igarashi 1988) were incorporated within a gas-permeable diffusion cell. Diffusion cells consisting of coiled Teflon™ tubing were first developed by Barber & Briegel (1987) to measure methane dissolved in groundwater. The oxygen sensors (diffusion cells) were tested and calibrated in the laboratory and response times were determined (Patterson et al. 1994). For the field trial, the oxygen sensors were bundled in groups of five and installed in backfilled boreholes across the water table and the NAPL-contaminated zone, at depths of 3.90, 4.15, 4.40, 4.65, and 5.15 m below ground. Oxygen concentrations were measured hourly over a period of 6 months using a specially designed multichannel data logger.

Cores were collected from each location where bundled oxygen sensors were emplaced. Details of the method of collection and analysis for NAPL content are given in Johnston & Patterson (1994). Analyses of the cores gave estimates of NAPL content prior to the remediation trial and enabled correlation of NAPL depth distributions with the depths of the oxygen sensors.

Total porosity and air-filled porosities were estimated from bulk density measurements and neutron moisture-meter logging within the trial area. Neutron-probe logging was carried out after the water tables were drawn down to approximately 4.5 m below ground, and involved lowering the probe down an access tube and recording count rates every 0.25 m to a depth of 5.5 m. The neutron-logged count rates are a direct measure of the liquid-filled (water and NAPL) porosity of the aquifer medium.

Rate coefficients for oxygen concentration decreases were estimated by fitting zero-order (linear decay) and first-order (exponential decay) curves to the oxygen depletion curves. Zero-order oxygen usage rate coefficients can be used to calculate NAPL degradation rates (Γ_h—mass NAPL degradation per mass of soil per day), using hexane as a model hydrocarbon, via the equation (e.g., Hinchee & Ong 1992):

$$\Gamma_h = (O_c \, P \, \theta_a)/(100 \, R \, T \, \rho_b) \cdot (MW_h)/(9.5) \tag{1}$$

where O_c is zero-order oxygen usage rate coefficient (%O_2/day), P is soil gas pressure (atm), R is the universal gas constant (8.206×10^{-5} m^3 atm/K/mole), T is temperature (K), θ_a is air-filled porosity (m^3/m^3), ρ_b is soil bulk density (kg/m^3), and MW_h is molecular weight of hexane (kg/mole).

RESULTS AND DISCUSSION

NAPL distributions at locations MI1 and MI3, prior to commencement of the remediation trial, are typical of those described by Johnston & Patterson (1994). As seen in Figure 1, the bulk of the NAPL was situated above the water table at the time of coring. NAPL is distributed over similar depth intervals (0.5 to 0.71 m thick) at locations MI1 and MI3, with the first occurrence of NAPL contamination 3.86 m below ground. Peak NAPL contents differ at the two locations; however, the total NAPL mass in the soil cores is similar.

Total soil-porosity estimates (Figure 2) are reasonably uniform throughout the soil profile, in the range 0.4 to 0.5. Air-filled porosity (θ_a) estimates

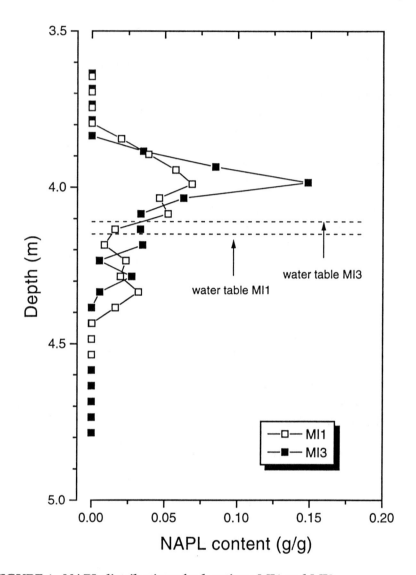

FIGURE 1. NAPL distributions for locations MI1 and MI3.

(Figure 2) decrease sharply with depth immediately above the water table. With respect to the water table position, θ_a profiles (depicted for the two dates in Figure 2) differ due to layering in the profile and hysteresis. Direct measurements are therefore critical to a good estimate of θ_a and, hence, NAPL degradation rate estimates. Fluctuating water tables, in particular, will impact θ_a (and oxygen concentration) measurements and, therefore, calculations of NAPL degradation rates.

During drawdown of water tables and aeration, a sharp decrease in θ_a down to 0.16 is observed at 3.75 m below ground, coinciding with a finer sand

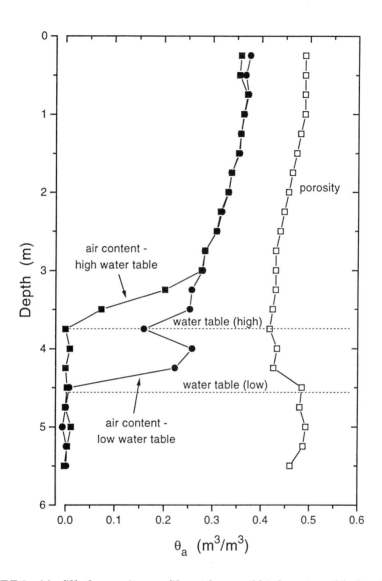

FIGURE 2. Air-filled porosity profiles at low and high water table levels, and the total aquifer porosity profile.

layer at that depth. The θ_a estimates are between 0.22 and 0.26 m³/m³ over the depth interval 4.0 to 4.25 m below ground, which is the top of the zone of NAPL contamination. Assuming a linear change in θ_a over the depths of the shallowest oxygen sensors, gives θ_a estimates of 0.22 m³/m³ for the 3.9-m depth and 0.24 m³/m³ for the 4.15-m depth.

Typical oxygen depletion profiles measured with the in situ oxygen sensor installations are shown in Figure 3. Note that although water tables were lowered to ~4.5 m below ground, the oxygen sensors at 4.4 m showed no response

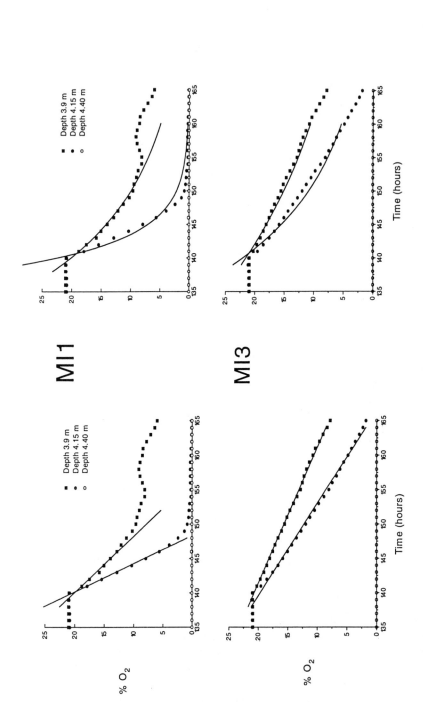

FIGURE 3. Linear (zero-order) and exponential (first-order) curves fitted to oxygen concentration depletion curves for locations MI1 (nutrient-amended) and MI3 (unamended). Aeration ceased at 140 h.

to aeration, indicating that water and/or NAPL had not drained below that depth. Groundwater samples taken at the same time and analyzed for dissolved oxygen also showed very low concentrations (< 0.5 mg/L).

Zero-order and first-order oxygen usage rate coefficients estimated from fitted curves in Figure 3 are collated in Table 1. The oxygen depletion curves for MI3 (unamended) and at 3.9 m for MI1 (nutrient amended) conform more closely to a straight line (zero-order) fit than an exponential (first-order) fit, although diffusion may also play a role in shaping the curves. Oxygen rate coefficients at the 4.15-m depth are greater than at the 3.9-m depth, which may be expected if θ_a were lower at deeper depths (i.e., nearer the water table), although measurements had shown increases with depth. It should be noted, however, that measurements of θ_a are very sensitive to the location of the water table and the variability of layering at the site. Increased rate coefficients at the 4.15-m depth could also be due to increased microbial activity within the NAPL-contaminated zone. Additionally, the estimates in Table 1 assume no transport of oxygen to the measurement location (i.e., a closed system). Preliminary modeling of the transport of oxygen by diffusion from the ground surface (i.e., an open system) suggests that the rates quoted in Table 1 may be underestimated by approximately 40%. This is to be further investigated.

From Equation 1, NAPL degradation rates between 7.8 and 40 mg-hexane/ kg-soil/day were calculated from the zero-order rate coefficients (Table 1). These rates are higher than, but comparable to, rates reported by Hinchee & Ong (1992), and substantially higher than rates estimated from data of Pijls et al. (1994) for fresh diesel contamination. The higher rates may be due to higher background rates of oxygen utilization or higher abiotic oxygen usage. Abiotic oxygen usage is thought to be low since only trace concentrations of readily oxidizable chemical species (e.g., iron and manganese) were measured in groundwater at the site. A further aeration test to determine background oxygen usage rates in an area uncontaminated with NAPL is currently being conducted.

TABLE 1. Zero-order and first-order oxygen usage rate coefficients, and hydrocarbon degradation rates calculated from the zero-order rates.

	Depth (m)	Zero-order oxygen use rate coefficients (%O_2/hour)	First-order oxygen use rate coefficients (hours^{-1})	Air-filled porosity (θ_a)	Degradation rates calculated from zero-order oxygen usage rate coefficients (mg-hexane/kg soil/day)
(MI1	3.90	1.2	7.1×10^{-2}	0.22	19
(Nutrient-	4.15	2.4	2.3×10^{-1}	0.24	40
Amended Area)					
MI3	3.90	0.51	3.6×10^{-2}	0.22	7.8
(Control Area)	4.15	0.74	7.1×10^{-2}	0.24	12

NAPL degradation rates determined for the nutrient-amended area of the trial plot were two to three times those in the unamended area, consistent with the laboratory observation that addition of nutrients enhanced biodegradation of the weathered diesel NAPL at the site. Pijls et al. (1994) reported a contrast in rate of between 5 and 8 for nutrient-amended and unamended sites of fresh diesel contamination, although no direct measure of porosity differences was indicated. No substantial reduction in NAPL content was observed when comparing cores recovered prior to the start of the trial and those recovered toward the end of the trial. However, at these carbon utilization rates, a NAPL content of 0.1 g/g would be reduced by only 5% over the 6-month trial.

SUMMARY AND CONCLUSIONS

In situ oxygen sensors have proved reliable and efficient for monitoring the progress of an in situ bioremediation field trial for weathered diesel NAPL contamination. The in situ sensors have supplied detailed vertical profile data on oxygen concentration changes in the vadose zone and groundwater on an hourly basis during the 6-month trial. Once calibrated, the oxygen sensor devices were as accurate as conventional oxygen monitoring techniques and required no maintenance or recalibration for a period of 6 months in conditions of varying water tables and NAPL contamination. The sensors were successfully linked to a data logger to provide semicontinuous data on oxygen concentrations.

Air-filled porosity estimates are sensitive to the location of the water table. Accurate measurement of the air-filled porosity, at the time and depth of measurement of the oxygen concentration decreases, has been shown to be critical for accurate estimation of NAPL degradation rates. NAPL degradation rates, calculated from oxygen depletion curves at two locations at the field site, show increased rates with depth. This may be due to different microbial communities, porosities, or NAPL concentrations at different depths within the soil profile. It may also be due to the replenishment of oxygen within the soil profile via diffusion from the ground surface.

NAPL degradation rate estimates calculated from the oxygen depletion curves are higher but comparable to rates reported by others, and indicate that enhanced biodegradation of weathered diesel NAPL has occurred in the field trial.

ACKNOWLEDGMENTS

Terry Power helped carry out laboratory testing of the oxygen sensors; Jack Smith constructed the multichannel logger; and David Briegel, Michael Lambert, Tracy Milligan, and Wayne Hick constructed and installed the multilevel oxygen sensors and operated the field trial.

REFERENCES

Barber, C., and D. Briegel. 1987. "A Method for the In Situ Determination of Dissolved Methane in Groundwater in Shallow Aquifers." *J. Contam. Hydrol.* 2: 51-60.

Hinchee, R.E., and S.K. Ong. 1992. "A Rapid In Situ Respiration Test for Measuring Aerobic Biodegradation Rates of Hydrocarbons in Soil." *J. Air Waste Manag. Assoc.* 42(10): 1305-1312.

Johnston, C.D., and B.M. Patterson. 1994. "Distribution of Nonaqueous Phase Liquid in a Layered Sandy Aquifer." In R.E. Hinchee, B.C. Alleman, R.E. Hoeppel, and R.N. Miller (Eds.), *Hydrocarbon Remediation*, pp. 431-437. Lewis Publishers, Boca Raton, FL.

Patterson, B.M., T.R. Power, G.B. Davis, and C. Barber. 1994. *An In Situ Device to Measure Oxygen in the Vadose Zone and Groundwater to Determine Oxygen Utilisation Rates: Laboratory and Field Evaluation.* CSIRO Division of Water Resources technical report No. 95/0.

Pijls, C.G.J.M., L.G.C.M. Urlings, H.B.R.J. van Vree, and F. Spuij. 1994. "Applications of In Situ Soil Vapor Extraction and Air Injection." In R.E. Hinchee (Ed.), *Air Sparging for Site Remediation*, pp. 128-136. Lewis Publishers, Boca Raton, FL.

Takeuchi, T., and I. Igarashi. 1988. "Limiting Current Type Oxygen Sensor." In T. Seiyama (Ed.), *Chemical Sensor Technology*, Vol 1, pp. 79-95. Kodansha Scientific, Japan.

Wood, B.D., C.K. Keller, and D.L. Johnstone. 1993. "In Situ Measurement of Microbial Activity and Controls on Microbial CO_2 Production in the Unsaturated Zone." *Wat. Resour. Res.* 29: 647-659.

Protocol Development for Determining Kinetics of In Situ Bioremediation

Henry H. Tabak, Rakesh Govind, Steven Pfanstiel,
Chunsheng Fu, Xuesheng Yan, and Chao Gao

ABSTRACT

Laboratory studies to determine biodegradation rates can be used as screening tests to determine the rate and extent of bioremediation that might be attained during remediation, and to provide design criteria (Graves et al. 1991; Govind et al. 1993). In this paper, a systematic protocol based on three types of bioreactors and one representative contaminant, phenol, has been developed to determine the biokinetic parameters of the suspended and immobilized microbiota and the transport parameters of contaminant and oxygen in the soil matrix. In the soil slurry reactor, significant degradation of contaminant occurs in the aqueous phase by the suspended soil microorganisms rather than by the soil immobilized biofilms. The soil slurry reactor data were used to derive the biokinetic parameters for the suspended and immobilized microorganisms. The wafer reactor data were used to obtain additional information with no oxygen limitations, and the porous tube reactor data provided a quantitative estimation of oxygen diffusivity in the soil matrix.

INTRODUCTION

Knowledge of biodegradation kinetics in soil is needed to understand the efficacy of in situ and ex situ bioremediation technologies. Laboratory studies to determine biodegradation rates can be used as screening tests to determine the rate and extent of bioremediation that might be attained during remediation, and to provide design criteria. Biodegradation in soil is a fairly complex process that involves diffusion of contaminants in the porous soil matrix, adsorption to the soil surface, biodegradation in the biofilms existing on the soil particle surface and in the large pores as well as in the bound and free water phase after desorption from the soil surface.

In this study, three types of bioreactors, shown in Figure 1, were used to determine the biokinetic parameters of the suspended and immobilized micro-

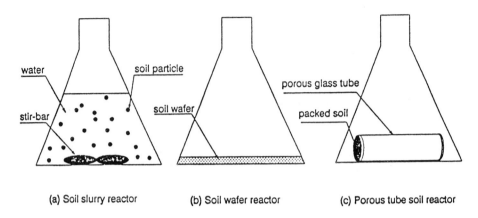

FIGURE 1. Schematic of soil slurry, wafer, and porous tube reactors.

biota and the transport parameters of contaminant and oxygen in the soil matrix: (1) the slurry bioreactor, where soil at 5% slurry concentration is vigorously mixed with the contaminant dissolved in water with nutrients; (2) the wafer reactor, where a thin wafer of soil is spiked with contaminant and nutrients dissolved in water, to obtain a 50% total soil moisture content; and (3) the porous glass tube reactor, where sieved soil with contaminant is packed in a porous glass tube with moisture content identical to the wafer reactor. In the soil slurry reactor there are no limitations of oxygen, which freely diffuses into the well-stirred slurry. Hence, the biodegradation rate in soil slurry reactors depends on the intrinsic biokinetic rate, microorganism concentration in the soil matrix, and inherent diffusivity of the contaminant. In the soil wafer reactor, oxygen diffuses freely through the thin soil matrix, and hence biodegradation rate is controlled by the water content in addition to other intrinsic parameters, as in the case of the soil slurry reactor. In the porous tube reactor, biodegradation rate is controlled by the water content and oxygen diffusivity and other intrinsic biokinetic parameters. The porous tube reactor provides a better estimate of biodegradation rates for in situ bioremediation than do the soil wafer and soil slurry reactors.

EXPERIMENTAL PROCEDURES AND MATERIALS

A Voith electrolytic respirometer was used to measure the cumulative oxygen uptake. First, 20 g of soil was mixed with 250 mL of distilled deionized (DD) water containing Organisation for Economic Cooperation and Development (OECD) nutrients and phenol at different concentration levels: 2.5, 5, and 10 mL of 2.5 g/L phenol stock solution equivalent to 6.25, 12.5, and 25 mg of phenol in each reactor. The nutrient concentrations as recommended by

OECD are (KH_2PO_4 (85 mg/L), K_2HPO_4 (217.5 mg/L)), $Na_2HPO_4.2H_2O$ (334 mg/L), NH_4Cl (25 mg/L), $MgSO_4.7H_2O$ (22.5 mg/L), $CaCl_2$ (27.5 mg/L) and $FeCl_3.6H_2O$ (0.25 mg/L), $MnSO_4.H_2O$ (0.0399 mg/L), H_3BO_3 (0.0572 mg/L), $ZnSO_4.7H_2O$ (0.0428 mg/L), $(NH_4)_6Mo_7O_{24}$ (0.0347 mg/L), $FeCl_3.EDTA$ (0.1 mg/L), and yeast extract (0.15 mg/L) (OECD 1981, 1983). Stock solution containing nutrients with no phenol was used for control flasks. The soil selected for this study was uncontaminated soil with the following characteristics: soil moisture 17%, organic matter 2.9%, classification silt loam, cation exchange capacity 6.5, soil pH 6.1, bulk density 1.06, nutrients in soil (ppm) phosphorus 17, potassium 90, magnesium 80, calcium 1100, and sodium 17. To determine the adsorption and desorption characteristics, the soil was air-dried and passed through a 2-mm sieve.

Soil Slurry Experiments

Specific procedures used in this research are presented as follows: (1) 20.0 g of soil was placed in a reaction flask, along with 250 mg of DD water and a Teflon™-coated stir bar; (2) Using a Teflon™ 3.0-mL-capacity syringe, 10.0 mL of 2.5-g/L phenol stock solution equivalent to 25 mg of phenol was transferred to the flask; and (3) soda lime was placed in the CO_2 trap, the flask was closed, and the respirometer run was started.

Soil Wafer Experiments

To more closely simulate the actual biodegradation of phenol in intact soils, soil wafer studies were developed for respirometer experimentation. The wafer system consisted of a thin layer (or wafer) of soil with a known moisture content. Procedures for the wafer experiments were as follows: (1) 20.0 g of soil and a measured amount of water, approximately 20 to 30 mL in volume, were placed in a reaction flask, well mixed to give uniform initial biomass concentration, and the entire flask was weighed; (2) the flask(s) were placed in the fume hood overnight and reweighed with the weighing procedure repeated until the weight of the flask indicated that the soil contained the desired moisture content; and (3) the soil wafer was contaminated with 10.0 mL of 2.5 g/L phenol stock solution using a syringe, followed by steps 3 and 4 for the slurry experiments. Due to the small wafer thickness (< 1 mm), diffusion of oxygen into the soil was not the controlling factor, resulting in a uniform oxygen concentration profile throughout the soil wafer.

Microporous Tube Experiments

Microporous tube reactors were developed for use in the electrolytic respirometer to study the effects of oxygen diffusivity. The porous glass tubes were purchased from Corning (Corning, New York) and were made from Vycor™ glass with pores averaging 40 angstroms (Å) in diameter. This pore size was selected because it was found to be optimum for containing the soil and water within the tube and allowing free flow of oxygen into the soil. The porous glass tube or pore sizes exceeding 40 Å were found not to affect the oxygen uptake results. The experimental procedure was as follows: (1) the microporous glass tubes were placed in the laboratory oven set at 200°C and left overnight to evaporate any water or contaminant in the pores or on the surface of the tubes; (2) the tubes were removed from the oven, allowed to cool for several hours at room temperature, and placed in a beaker containing DD water to initially saturate the pores in the glass; (3) the tubes were removed from the water and filled with 20.0 g of soil, plugging both ends with glass wool; (4) the soil was contaminated evenly in the tube with 1.25 to 5.0 mL of experimental stock by inserting a syringe through the plugs at either end and mixing the soil with the syringe as much as possible while injecting the solution; and (5) two tubes were placed into each flask, followed by steps 3 and 4 for the slurry experiments.

Adsorption and Desorption Experiments

For adsorption experiments, 50-g soil samples were placed in a 250-mL glass bottles. Various concentrations for each compound, namely 10, 25, 50, 100, and 150 mg/L, were used. The volume of the total solution added to each bottle was 250 mL to maintain a minimum headspace. The headspace was important, as it not only affected the degree of mixing during stirring but also controlled the loss of phenol due to volatilization. Blanks containing only the phenol solution allowed the measurement of volatilization losses. To prevent phenol degradation in solution, 1 mL $HgCl_2$ saturated solution was added to each bottle. All adsorption experiments were conducted in a fume hood, so that the temperature of adsorption could be controlled at 24°C.

The mixture in the bottle was stirred using a magnetic stirrer for 96 h. The samples at 2, 4, 6, 8, 12, 24, 36, 48, 72, and 96 h from the beginning of the adsorption experiment were filtered using a 0.45-μm silver membrane and placed in a 50-mL sample vial for extraction. The extraction method was based on U.S. Environmental Protection Agency Methods 604 and 610.

The GC was an HP-5890A model equipped with a flame ionization detector. The following conditions were used throughout this study: initial oven temperature 60°C, initial hold time 5 min, oven temperature rate 8°C/min., final oven temperature 280°C, final hold time 5 min, injector temperature 225°C, detector temperature 310°C, makeup gas (nitrogen) flowrate 35 mL/min, detector gas flowrate 32 mL/min hydrogen and 435 mL/min air, carrier gas (helium) 2 mL/min, column HP-5 methyl silicon gum and 5 m × 1.53 mm × 2.54 mm film thickness, and software HP Chemstation.

After adsorption equilibrium was attained, the solution in a 250-mL glass bottle was mixed with 250 mL DD water in a large 500-mL glass bottle to start the desorption experiment. All the conditions for the desorption experiment were the same as for the adsorption experiment. Samples for quantifying desorption were treated the same way as for adsorption studies, although the time taken to achieve desorption equilibrium was significantly longer.

RESULTS AND DISCUSSION

Detailed mathematical models were developed for the slurry, wafer, and porous tube reactors. The following soil parameters were determined through separate experiments: (1) soil particle size distribution was determined using methods outlined by Day (1965); (2) soil porosity, pore size distribution, pore volume, and surface area were determined by nitrogen adsorption using Micrometrics ASAP 200, the Brunauer, Emmett, and Teller (BET) specific surface area $(m^2/g) = 20.27$, BET void volume $(cm^3/g) = 0.029$, BET average pore diameter $(\text{Å}) = 58.0$, and soil porosity $= 0.06525$; and (3) Freundlich isotherms were used to quantitate soil adsorption/desorption isotherms.

Figure 2 shows the oxygen uptake data for each of the experimental schemes for 100 mg/L phenol concentration. The oxygen uptake curve for the slurry case attains a higher plateau than for the soil wafer and porous tube reactors. This indicates that more phenol degraded in the slurry reactor than in the wafer and porous tube reactors. This is mainly attributed to low water content in the wafer and porous tube reactors, thereby limiting biodegradation in the bound and free water phase by the suspended and immobilized microorganisms. In the case of

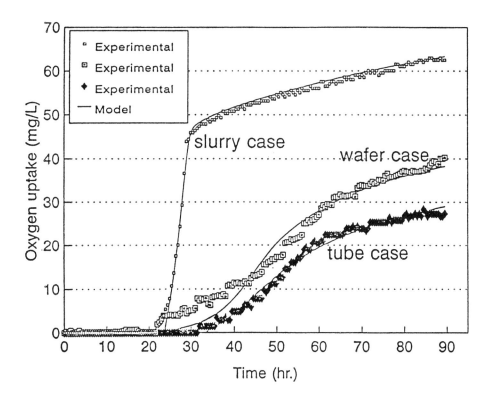

FIGURE 2. Cumulative oxygen uptake data and model fits for soil slurry, wafer, and porous tube reactors.

the soil slurry reactor, the extent of biodegradation in the slurry water phase due to suspended microbiota is significant. The extent of biodegradation in the wafer exceeded that in the porous tube reactor mainly due to oxygen limitations in the porous tube soil. The rate of oxygen uptake in the slurry reactor was also the highest when compared to the other reactors. The soil slurry reactor attains a plateau value in about 1 day, whereas the wafer and tube reactor's cumulative oxygen uptake attains a plateau value in about 3.5 days.

The oxygen uptake data for the soil slurry, wafer, and tube reactors were analyzed using a detailed mathematical model to determine the best-fit values for the model parameters. The model is based on diffusion and reaction equations using the Monod model for biodegradation in the soil and aqueous phases. The soil slurry reactor was used to determine the biokinetic parameters for the suspended and soil immobilized microbiota. These parameters, when used with the appropriate amount of free water, were used to fit the wafer reactor data. The model did not closely fit the initial oxygen uptake data for the wafer reactor due to variabilities in phenol concentration in the soil matrix. The porous tube reactor data were used to derive the oxygen diffusivity in the soil phase. The model fits are also shown in Figure 2 along with the experimental data. Determining the oxygen profile in the porous tube soil using the model showed that the radial oxygen concentration decreases rapidly, attaining a zero value at a radial distance of 0.25R from the tube center, where R is the radius of the porous tube. This again confirmed that there were oxygen limitations in the porous tube reactor due to limited oxygen diffusion in the soil matrix.

Experiments with uniformly labeled C^{14} phenol and measurement of carbon dioxide evolution by absorbing in KOH solution and scintillation counting showed that the net oxygen uptake (actual uptake minus the oxygen uptake in the control flask) was due solely to phenol degradation. This verified our initial assumption that the net cumulative oxygen uptake in each type of soil reactor could be used to derive the biokinetic and transport parameters.

DESCRIPTION OF PROTOCOL

The overall protocol for evaluating soil biodegradation kinetics involves the following steps:

1. Measure cumulative oxygen uptake in soil slurry reactors. The slurry reactor model is used to obtain the aqueous- and soil-phase Monod biokinetic parameters.
2. Measure cumulative oxygen uptake in soil wafer reactor. Use the biokinetic parameters determined from the slurry model and detailed wafer model to determine the diffusivity of the contaminant in the soil matrix.

3. Measure cumulative oxygen uptake in porous tube reactor. Use the parameters determined from the soil slurry and wafer models to calculate oxygen diffusivity in the soil from the porous tube reactor model.
4. Measure radiolabeled carbon dioxide evolution using a uniformly labeled compound in soil slurry, wafer, and porous tube reactors to verify that oxygen uptake was due to compound mineralization. The biokinetic parameters, compound, and oxygen diffusivity in soil determined using the protocol can be used to model the kinetics of in situ bioremediation.

REFERENCES

Day, P.R. 1965. "Particle Fractionation and Particle-Size Analysis." In C.A. Black (Ed.), *Methods of Soil Analysis*, 43, American Society of Agronomy, Madison, WI.

Graves, D.A., C.A. Lang, and M.E. Leavitt. 1991. "Respirometric Analysis of the Biodegradation of Organic Contaminants in Soil and Water." *Applied Biochem. Biotech. 28/29*: 813-826.

Govind, R., X. Yan, L. Lai, S. Pfanstiel, C. Gao, and H.H. Tabak. 1993. "Development of Methodology to Determine the Bioavailability and Biodegradation Kinetics of Toxic Organic Pollutant Compounds in Soil." In R.E. Hinchee, D.B. Anderson, F.B. Metting, Jr., and G.D. Sayles (Eds.), *Applied Biotechnology for Site Remediation*, pp. 229-239, Lewis Publishers, Boca Raton, FL.

OECD. 1983. *OECD Guidelines for Testing of Chemicals*, EEC Directive 79/831, Annex V, Part C: Methods for Determination of Ecotoxicity, 5.2 Degradation. Biotic Degradation. Manometric Respirometry. Method DGXI, Revision 5.

OECD. 1981. *OECD Guidelines for Testing of Chemicals*, Section 3, Degradation and Accumulation, Method 301C, Ready Biodegradability: Modified MITI Test (I) adopted May 12, 1981 and Method 302C Inherent Biodegradability: Modified MITI Test (II) adopted May 12, 1981, Director of Information, OECD, Paris, France.

Vadose Zone Nonaqueous-Phase Liquid Characterization Using Partitioning Gas Tracers

G. Allen Whitley, Jr., Gary A. Pope,
Daene C. McKinney, Bruce A. Rouse,
and Paul E. Mariner

ABSTRACT

Researchers of various disciplines associated with environmental cleanup are beginning to investigate the use of partitioning tracers to determine and characterize contaminated sites. Characterizing the vadose, or unsaturated zone, as well as the saturated zone is imperative because there are areas in the western United States with large vadose zones that are contaminated by nonaqueous-phase liquids (NAPLs). This paper presents laboratory experiments conducted to determine the partition coefficients between air and NAPL for several gas tracers. Once the tracers' partition coefficients have been estimated, they may be used to estimate NAPL volumes in situ. The experiments entailed the introduction of trichloroethylene (TCE) into a column packed with Ottawa sand and containing a residual amount of water. With a known amount of TCE in the column, partitioning gas tracers were injected and their breakthrough monitored. Using the method of moments, the partition coefficients of the tracers were estimated. The gas tracers used in this experiment included argon as a nonpartitioning tracer and several perfluorocarbons as partitioning tracers. The experimental results demonstrate that partitioning gas tracers and, in particular, perfluorocarbons, may be used to estimate residual NAPL saturations in the vadose zone.

INTRODUCTION

A widespread and recurring public health problem surfacing in recent years is the contamination of the subsurface by NAPLs. The typical scenario for their infiltration into the subsurface entails a waste disposal site that has received wastes from various operations for a number of years. These wastes typically contain NAPLs that may establish sufficient hydraulic potential in their receptacles, usually pits dug in the ground, to infiltrate the surface and make their way through the vadose zone to the water table.

During this infiltration process, the NAPLs, both light and dense, leave a tail or streak through the vadose zone. This trail, held in place by capillary forces, is defined as residual or immobile NAPL. In some locales, this NAPL residual is the only contamination, and the hydraulic potential is not great enough for the NAPL to reach the water table. The characterization of this residual NAPL volume is the focus of this paper.

Partitioning tracer tests have been used to characterize liquid residuals in oil and gas reservoirs (Allison et al. 1991; Tang 1992). In a partitioning interwell tracer test (PITT), partitioning tracers experience a delay in production with respect to nonpartitioning tracers due to their interaction with the resident oil. If the partition coefficient between the carrier fluid and the oil is known, the difference between the breakthrough curves of the partitioning and nonpartitioning tracers can be used to estimate the volume of oil present. Using Henry's law, chromatographic theory, and various partitioning gas tracers, oil saturations were estimated in slim tube displacement experiments and field tests (Tang and Harker 1991a,b).

Recently, the PITT has been adapted to the detection and characterization of NAPL contamination in the subsurface. This procedure has been successfully demonstrated in saturated column experiments and 2- and 3-dimensional computational experiments with dense, nonaqueous-phase liquids (DNAPLs) (Jin et al. 1995).

Given the results of a PITT, the method of moments provides a method for NAPL characterization in the vadose zone. If the partition coefficient of the tracer between gas and NAPL is known, the method of moments can be used to estimate the NAPL saturation based on the gas tracer response curves.

The tracers chosen for this study consist of the perfluorocarbon family; a family of fully fluorinated alkyl-substituted alkanes. Perfluorocarbon tracers (PFTs) were studied previously for petroleum reservoir characterization by Senum et al. (1992). Although there are many compounds in this group, the ones considered in this study are listed in Table 1 along with their molecular weights, vapor pressures, and boiling temperatures. Another tracer of interest, because of its potential for partitioning into NAPLs and its previous use in tracer studies (Tang and Harker 1991b), is sulfur hexafluoride. The nonpartitioning tracer used in this study was argon.

THEORY

Method of Moments

A detailed description of the method of moments is given by Jin et al. (1995) to derive the following equation:

$$V_N = \frac{q(\bar{t}_2 - \bar{t}_1)}{K_2 - K_1} \tag{1}$$

TABLE 1. Properties of perfluorocarbons used in partitioning tracer experiments.

Tracer	Molecular Weight (g/mole)	Vapor Pressure (psi @ °C)	Boiling Point (°C)
Argon, Ar	39.94	—	−186.0
Sulfur Hexafluoride, SF_6	145.06	334 @ 21	−63.8
Carbon Tetrafluoride, CF_4	88.01	—	−128.0
Octafluorocyclobutane, C_4F_8	200.04	40 @ 21	−5.8
Octafluorocyclopentene, C_5F_8	212.05	12.7 @ 25	27.0
Dodecafluorodimethylcyclobutane, C_6F_{12}	300.06	7.2 @ 25	45.0

where V_N is the volume of NAPL [L^3] in the porous medium, q is the flowrate [L^3/T], $\bar{t}_1$ and $\bar{t}_2$ are the mean residence times [T] of the response curves of tracers 1 and 2 as computed from the first moments of the data, and K_i is the partition coefficient of tracer i between air and NAPL, the ratio of its concentration [M/L^3] in the NAPL phase to its concentration in the air phase

$$K_i = \frac{C_{iN}}{C_{ia}} \qquad (2)$$

Additionally, $\bar{t}_i$ is defined as

$$\bar{t}_i = \frac{\int_0^\infty C_{ia}t\,dt}{\int_0^\infty C_{ia}\,dt} - \frac{t_{is}}{2} \qquad (3)$$

where t_{is} is the slug size of tracer i.

It is common in a PITT to use a nonpartitioning tracer ($K_i = 0$) as a reference tracer (one that will not be delayed due to the presence of NAPL) to determine the unretarded tracer response. When a nonpartitioning tracer is used, Eq. 1 can be simplified to

$$V_N = \frac{q(\bar{t}_p - \bar{t}_n)}{K_i} \qquad (4)$$

where K_i is the partition coefficient of partitioning tracer i between the NAPL and air, and $\bar{t}_p$ and $\bar{t}_n$ are the mean residence times of the partitioning and non-partitioning tracers in the porous medium, obtained by integrating the two tracer response curves to estimate the first moments of the data. This equation can be used to evaluate the volume of NAPL residing between two wells, or, as in our case, given V_N measured in the laboratory, the partition coefficient K_i may be calculated.

MATERIALS AND METHODS

Materials

Table 1 lists some of the properties of the selected PFTs, as well as those of argon and sulfur hexafluoride. The NAPL used in the experiments was TCE. The density of TCE at 20°C is 1.462 g/cc (Reid et al. 1987). Its molecular weight is 131.39 g/mole, and its vapor pressure is 1.16 psi at 20°C. Ottawa sand with a size range from 40 to 140 U.S. sieve numbers or 0.425 to 0.105 mm was used in 2.5-cm columns.

Column Procedures

Three columns were prepared for each experiment: (1) a contaminated primary column (2.5 cm × 30 cm) containing residual water and TCE in which the partitioning tracer studies occur; (2) a contaminated precolumn (2.5 cm × 60 cm) containing water and TCE used to saturate incoming air with water and TCE so that the air and TCE will not be stripped from the primary column; and (3) an uncontaminated secondary column (2.5 cm × 30 cm) containing only water used to examine tracer response in a clean column.

All three columns were packed with sand on a vibrating jig. The columns were leakchecked at 10 psi (68,947 Pa) and weighed before and after packing to determine porosity and pore volume. The porosity and pore volume of the primary column were estimated to be 33.6% and 54.6 cc, respectively.

The primary column was saturated with deionized water, which was gravity-drained into the column. The difference between the water injected and the water produced was used to determine the residual water saturation. The column was weighed to verify the water retained. The residual water saturation was estimated to be 35.9%.

Next, approximately 5 cc of TCE, enough to produce a residual saturation, was injected into the primary column and allowed to gravity drain. The column was weighed for a more accurate estimate of TCE saturation. The residual TCE saturation was estimated to be 8.4%. In both cases, water and TCE injection, the precolumn was treated in the same manner except that larger amounts were introduced into the precolumn. Figure 1 illustrates the experimental setup.

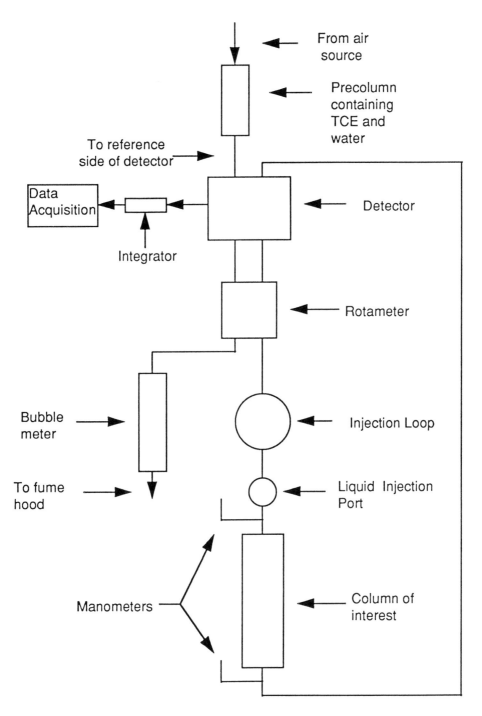

FIGURE 1. Experimental apparatus for gas tracer studies.

The detection method used for this experiment was a thermal conductivity detector connected in line with the output of the column. The output signal was monitored continuously on a linear chart recorder, and signals were sent once per minute to a data acquisition system.

Injection Method

The injection procedure was as follows: airflow was established and a baseline on the chart recorder was established. The air was saturated with water and TCE by passing it through the precolumn. Next, the tracers were injected sequentially. The tracers that are gases at experimental temperature and pressure (argon, CF_4, C_4F_8) were injected through a 2.6-cc gas chromatograph (GC) injection loop. The tracers that are liquids at the same conditions (C_5F_8, C_6F_{12}) were injected through a GC injection port heated to 40°C. The liquid volume of C_6F_{12} injected was 0.1 cc and the liquid volume of C_5F_8 injected was 0.05 cc. The liquids vaporize as they mix with the air stream and then flow through the column as gas mixtures.

RESULTS AND DISCUSSION

Uncontaminated Column Results

Argon and C_6F_{12} were injected into the uncontaminated column to monitor the C_6F_{12} with respect to argon. The mean residence times were calculated by the method of moments to be 45 cc for both the argon and C_6F_{12} tracers. This result shows that there was no partitioning or adsorption of the C_6F_{12} in this clean column, because it did not separate from the inert argon tracer. Thus, we are justified in interpreting the results given below showing separation of these tracers in the columns contaminated with TCE as entirely due to the partitioning into the TCE residual phase. Under the conditions of these experiments, this residual phase is present as a separate liquid phase trapped by capillary forces in the form of ganglia rather than as adsorbed organic on the sand particles.

Contaminated Column Results

The mean residence time for production of each tracer was calculated from the response curves using the method of moments, and Eq. 2 was used to calculate the experimental partition coefficient for each tracer. Table 2 and Figure 2 provide tabular and graphical analysis of the results at a flowrate of 0.45 cc/min.

Table 2 lists the mean residence time (in min and cc) and experimental partition coefficients calculated for each tracer. Argon has the shortest mean residence time, followed by CF_4 and SF_6. The small retention time for CF_4 indicates

TABLE 2.   Partition coefficients from partitioning tracer experiment at 0.45 cc/min with Ottawa sand contaminated with TCE.

Tracer *i*	Mean Residence Times		K_i (C_{TCE}/C_{AIR})
	$\bar{t}_i$ (min)	$\bar{t}_i$ (cc)	
Argon	93.2	41.9	0.00
SF_6	103.7	46.7	1.02
CF_4	95.1	42.8	0.22
C_4F_8	131.4	59.1	3.73
C_5F_8	235.0	105.8	15.4
C_6F_{12}	187.6	84.4	9.0

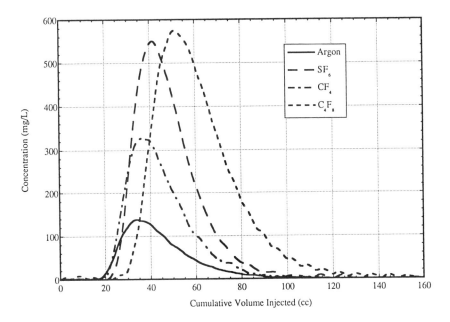

FIGURE 2.   Response curves for tracers in TCE-contaminated Ottawa sand at 0.45 cc/min.

that its partition coefficient is essentially zero because the value falls within the range of measurement error compared to argon. However, SF_6 does show a small but experimentally significant partitioning into the TCE.

Further experiments were conducted with flowrates of 1.38 cc/min and 4.15 cc/min. The same suite of tracers was injected sequentially into the column at these flowrates. These rates correspond to linear velocities of 2×10^{-4} m/s and 6.4×10^{-4} m/s. The resulting partition coefficients are listed in Table 3. These data indicate that equilibrium partitioning is not occurring at the higher flow rates. Because a rate dependence was observed, the rate was lowered to 0.08 cc/min and the experiment was repeated for C_4F_8. A batch equilibrium measurement of the partition coefficient was made for this tracer. Both of these measurements gave a partition coefficient of about 4.8, which is significantly higher than the value of 3.7 at the next lowest rate of 0.45 cc/min. Figure 3 is a plot of these data to illustrate the flowrate dependence. Using the partition coefficient for C_4F_8 of 4.8 as the equilibrium value gives an activity coefficient of 26.6. This indicates that the C_4F_8 in TCE-rich organic liquid phase (the NAPL) is significantly nonideal in the thermodynamic sense. Thus, using Raoult's law would not be a good approximation for estimating the partition coefficients.

Another experiment was conducted involving the injection of tracers into a column packed with uncontaminated field soil containing clays. The porosity of the field soil was 42.6%. The permeability was estimated to be about 1.3 darcys, which is about six times lower than that of the Ottawa sand. The soil was contaminated with TCE in the same manner as the previous experiments. The saturation of TCE estimated by weighing the column before and after the contamination with TCE was 0.23. The tracers were then injected at

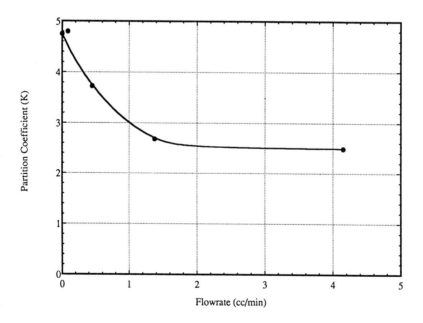

FIGURE 3. Partition coefficients versus flowrate for C_4F_8.

TABLE 3. Experimental partition coefficients (K_i) at various flowrates for Ottawa sand and field soil contaminated with TCE.

Tracer	K_i (C_{TCE}/C_{AIR}) at 1.38 cc/min Ottawa sand	K_i (C_{TCE}/C_{AIR}) at 4.15 cc/min Ottawa sand	K_i (C_{TCE}/C_{AIR}) at 0.4 cc/min Field soil
SF_6	0.04	0.37	0.69
CF_4	—	−0.21	0.12
C_4F_8	2.68	2.49	4.77
C_5F_8	13.8	12.8	—
C_6F_{12}	—	—	11.8

0.4 cc/min, corresponding to a linear velocity of 5.8×10^{-5} m/s. The estimated partition coefficients for the various tracers in field soil are given in Table 3. The CF_4 again shows negligible partitioning into the TCE. The other tracers show partitioning similar to the previous experiments. The partition coefficient of C_4F_8 was estimated to be 4.77, almost the same as the batch equilibrium measurement. Alternatively, if the partition coefficient of 4.8 measured from the batch equilibrium experiment is used and the saturation of TCE calculated from the retention times of the C_4F_8 and argon, then the estimated saturation is 0.23, in agreement with the mass balance from weighing the column. In general, we expect to be able to estimate saturation under these conditions to within about 10%.

CONCLUSIONS

Tracers that are suitable to inject with air as a carrier phase into the vadose zone of a TCE- contaminated waste site to estimate the amount of contaminant have been identified by performing column experiments. These perfluorocarbon tracers appear to be excellent choices as partitioning tracers, and either argon or carbon tetrafluoride should be suitable nonpartitioning tracers to use as a reference. Perfluorocarbon gas tracers used in partitioning interwell tracer tests are expected to be an effective means of NAPL contamination detection, characterization, and remedial action performance assessment within the vadose zone. Some specific conclusions drawn from this study based on several column experiments in Ottawa sand and a field soil are:

1. The higher the molecular weight of the perfluorocarbon and the lower its vapor pressure, the greater its partitioning into TCE.

2. The mixture between the perfluorocarbons and TCE is highly nonideal, as indicated by the large activity coefficient of C_4F_8 in TCE.
3. The partition coefficients calculated from column data decreased as the flowrate increased, which indicates that equilibrium partitioning is not occurring at the higher rates. Thus, column experiments must be done at a low flowrate to obtain equilibrium partition coefficients. Residence times of at least 10 h appear to be necessary under the specific conditions of these experiments, but in general the minimum residence time must be determined experimentally for the specific column of interest.
4. The perfluorocarbon tracers exhibited similar characteristics in field soil and Ottawa sand.
5. Adsorption of the perfluorocarbon tracers on the sand was negligible. Thus, the entire retardation of the tracers can be attributed to partitioning into the residual TCE phase rather than adsorption onto the soil.

There are of course many additional considerations to take into account before such tracers are used in the field. Although these are beyond the scope of this paper, some of them have been addressed in related papers by Jin et al. (1995) and Pope et al. (1995) and will be the subject of future papers as well. Briefly, we rely heavily on three-dimensional modeling of the contaminated field site to design partitioning tracer tests taking into account complications such as heterogeneity, unconfined flow, nonuniform distribution of NAPL, and so forth. An example of one such field design can be found in Pope et al. (1995). The results of the successful field test conducted during 1994 based on this design can be found in Annable et al. (1995).

ACKNOWLEDGMENTS

The authors would like to thank Richard E. Jackson and Cindy Ardito of Intera, Inc. for useful discussions and information about the application of partitioning tracers to contaminated vadose zones. Funding for this research was provided by Sandia National Laboratories, Contract No. AG-2475.

REFERENCES

Allison, S.B., G.A. Pope, and K. Sepehrnoori. 1991. "Analysis of Field Tracers for Reservoir Description." *J. Pet. Sci. Eng.*, 5: 173.
Annable, M.D., P.S.C. Rao, K. Hatfield, W.D. Graham, and A.L. Wood. 1995. "Use of Partitioning Tracers for Measuring Residual NAPL Distribution in a Contaminated Aquifer: Preliminary Results from a Field-Scale Test." Proceedings from the Second Tracer Workshop at The University of Texas at Austin, November 14-15, 1994, IFE/KR/E-95/002, Institute for Energy Technology, Kjeller, Norway.
Jin, M., M. Delshad, V. Dwarakanath, D.C. McKinney, G.A. Pope, K. Sepehrnoori, and C.E. Tilburg. 1995. "Partitioning Tracer Test for Detection, Estimation and Remediation

Performance Assessment of Subsurface Nonaqueous Phase Liquids." In press, *Water Resour. Res.*

Pope, G.A., M. Jin, V. Dwarakanath, B.A. Rouse, and K. Sepehrnoori. 1995. "Partitioning Tracer Tests to Characterize Organic Contaminants." Proceedings from the Second Tracer Workshop at The University of Texas at Austin, November 14-15, 1994, IFE/KR/E-95/002, Institute for Energy Technology, Kjeller, Norway.

Reid, R.C., J.M. Prausnitz, and B.E. Poling. 1987. *The Properties of Gases & Liquids.* 4th ed., McGraw-Hill, Inc., New York, NY.

Senum, G.I., R.W. Fajer, B.R. Harris, Jr., W.E. DeRose, and W.L. Ottaviani. 1992. "Petroleum Reservoir Characterization by Perfluorocarbon Tracers." Proceedings of the SPE/DOE Eighth Symposium on Enhanced Oil Recovery, Vol. 2, 337. Tulsa, OK.

Tang, J.S. 1992. "Interwell Tracer Test to Determine Residual Oil Saturation to Waterflood at Judy Creek BHL 'A' Pool." *J. Can. Pet. Tech., 31* (8): 61.

Tang, J.S. and B. Harker. 1991a. "Interwell Tracer Test to Determine Residual Oil Saturation in a Gas-Saturated Reservoirs. Part I: Theory and Design." *J. Can. Pet. Tech., 30* (3): 76.

Tang, J.S. and B. Harker. 1991b. "Interwell Tracer Test to Determine Residual Oil Saturation in a Gas-Saturated Reservoirs. Part II: Field Applications." *J. Can. Pet. Tech., 30* (4): 34.

Toxicity Screening of Waste Products Using Cell Culture Techniques

Marc Petitmermet, Alain Favre, Brigitte Shah,
Ursula Rösler, Jörg Mayer, and Erich Wintermantel

ABSTRACT

The cell reactions due to the presence of residues and their extracts were studied using quantitative and qualitative methods. The results of the applied cell culture techniques showed that fly ash was much more cytotoxic than slag. This finding correlates with the chemical analysis. The washed samples (H_2O, HCl, H_2SO_4, HNO_3) were again less cytotoxic than their corresponding unwashed samples due to the lack of water-soluble compounds. The very sensitive response of the cell cultures to toxic substances was used to classify and validate the applied treatment methods.

INTRODUCTION

More than 600,000 tons of residue from waste incineration plants is produced in Switzerland each year. These residues are slag, fly ashes, and residues from extended flue gas cleaning. Because they are contaminated with heavy metals, they have to be deposited in appropriate landfills. Due to the increasing amount of municipal and industrial waste and the decreasing amount of disposal sites, additional treatment of waste and its by-products is becoming more and more important. To decrease the amount of residuals to be deposited, the heavy metal content of the residues has to be reduced by physical, chemical, or biological methods to acceptably low levels to obtain products suitable for reuse in the construction industry. Possible approaches are: extended physical separation, chemical extraction of the heavy metals with acids, bioleaching of the heavy metals with microorganisms (Barrett et al. 1993; Burgstaller & Schinner 1993), and rendering hazardous substances inert and immobilizing them by vitrification.

To ensure environmental safety, the toxicity of the residues from different treatment methods has to be monitored. Chemical, physical, and biological analysis methods, e.g., microorganisms (Lahmann et al. 1989), fish cell

Monitoring and Verification of Bioremediation

cultures (Ahne 1985; Babich et al., 1986; Hansen 1989), or human cell cultures (Jellum et al. 1992; Sewing 1994) can be used. Cell culture techniques may be applied to measure acute cytotoxicity as a more sensitive method. This work will concentrate on the investigation of washed residues and vitrified samples.

MATERIALS AND METHODS

The materials investigated are presented in Table 1.

Cell Culturing

The 3T3 mouse fibroblasts (connective tissue cells) were cultivated in DMEM (Dulbecco's Modified Eagle Medium, Gibco) containing 10% FCS

TABLE 1. Description of test materials used, their abbreviations, treatments, and providers.

Material	Abbreviation	Treatment	Source
fly ash	FU	untreated	waste incineration plant, KEZO, Hinwil, Sulzer Chemtech, Winterthur, Switzerland
	FM	ground (∅ < 0.5 mm)	ditto
	FW	washed (H_2O)	ditto
	FU HCl	washed (HCl)	ditto
	FU H_2SO_4	washed (H_2SO_4)	ditto
	FU HNO_3	washed (HNO_3)	ditto
slag	SUS	untreated, fraction < 0.5 mm	ditto
	SWS	washed (H_2O), fraction < 0.5 mm	ditto
metal hydroxides	MHV	vitrified	pilot plant, Sulzer Chemtech, Winterthur, Switzerland
paint sludge	PSV	vitrified	
fly ash	FAV	vitrified	DEGLOR pilot plant, ABB, Baden-Dättwil, Switzerland
bottle glass	BGV	vitrified	—

(Foetal Calf Serum, Gibco) and 1% PS (Penicillin-Streptomycin, Gibco) at 37°C, 5% CO_2, and 95% relative humidity.

Qualitative Examinations Using
Scanning Electron Microscope (SEM)

After seeding fibroblast cells onto the surface of the test materials and incubating for 6 h up to 14 d, the protein structure of the cells was fixed with 5% paraformaldehyde in phosphate-buffered saline (PBS) for at least 1 h at 4°C. The specimen was dehydrated with increasing concentrations of ethyl alcohol and critical point drying with CO_2 (Critical Point Dryer, CPD 030, Bal-Tec), then were sputtered with platinum (500 s, 9 mA) and studied by scanning electron microscope (SEM, Hitachi S-2500C).

Semiquantitative Examinations Using
the Agar Diffusion Overlay Test
(ASTM 1984; ISO 1992)

After a cell monolayer on the culture dish was established, the cell culture medium was changed with a medium containing approximately 2% agar. The agar medium solidified and allowed specimens to be put onto the agar surface: solid samples as compact bodies, powder in tubes, and liquids by saturating filter disks. Soluble compounds diffused through the agar medium to the cell monolayer at the bottom of the culture dish and affected the cell activity. After 24 h, the cells were marked with neutral red agent (NR, Flucka). The neutral red agent diffused through the agar medium to the cell monolayer into the cells where it was actively accumulated in the lysosomes. Depending on the velocity of diffusion and the toxicity of the soluble compounds, cells accumulated more or less neutral red according to their activity. Therefore, the diameter of the zone of damaged or avital cells could be used to determine the cytotoxicity of the test specimen. Semiquantification was performed according to ASTM F895-84 (1984) to calculate response indices (Figure 1).

Quantitative Colorimetric Measurements
Using Direct Contact Methods

By using the appropriate test methods, the influence of toxic substances on cell functions can be quantified. The 3-(4,5)-dimethylthiazol-2-yl-2,5-diphenyl-tetrazoliumbromide (MTT) test measures the dehydrogenase activity of the cells depending on the concentration of added specimens (cell proliferation kit I). Samples with different final concentrations were put into 24-well plates, 20,000 cells/well were then added. After 2 days of incubation, the MTT agent was added to the medium. This agent diffuses to the mitochondria, where the dehydrogenase enzyme decomposes MTT to a formazan salt. This salt has an absorption maximum of 560 nm, which was measured with a 96-well plate

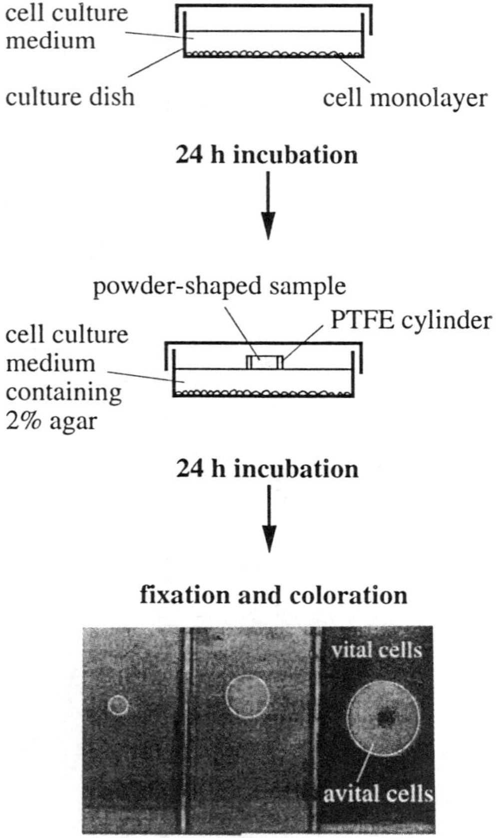

Response Index = Zone Index / Lysis Index

Zone Index (ZI)

Index	Description of zone
0	No detectable zone around or under sample
1	Zone limited to area under sample
2	Zone not greater than 5 mm in extension from sample
3	Zone not greater than 10 mm in extension from sample
4	Zone greater than 10 mm in extension from sample, but not involving entire plate
5	Zone involving entire plate

Lysis Index (LI)

Index	Description of zone
0	No observable lysis
1	Up to 20% of zone lysed
2	Over 20% to 40% of the zone lysed
3	Over 40% to 60% of the zone lysed
4	Over 60% to 80% of the zone lysed
5	Over 80% lysed within zone

FIGURE 1. Agar diffusion test (ASTM F895-84).

reader (Rainbow, Slt, Austria). The dehydrogenase activity can be correlated to the viability of the cells and, therefore, to the cytotoxicity of the samples.

Quantitative Colorimetric Measurements Using Indirect Methods (ISO 1992)

Instead of using the specimens directly, their extract can be added to the culture medium. Extraction was performed in a lab shaker at 37°C, 80 rpm/min for 3 days. As extraction medium, the cell culture medium itself was used according to ISO 10993-5 (1992). The 96-well plates were incubated with about 10,000 cells/well and 100 μL cell culture medium for 1 h. In the meantime, solid samples were aseptically removed from the extraction medium using centrifugation (Universal/K2S, Hettich, Germany) and microfiltration (0.2 μm) techniques. The pH value was then adjusted to about 7.4, and the dilution series was prepared using disposable 12-channel reservoirs. Culture medium was then exchanged with the extract-containing medium in a decreasing dilution series. After 2 days of incubation, the viability of the cells was measured with the MTT test as previously described.

RESULTS

Qualitative Examinations by SEM

A first qualitative comparison of treated and untreated residues is shown in Figures 2 and 3. Due to the toxicity and the local pH change, fibroblasts did not grow on untreated fly ash (FU) and water washed fly ash (FW). The lack of philapodiae (see Figure 3, P), the porous cell membrane, and the spherical shape are indications that the cells are avital (Figure 2, right).

On the other hand, fibroblasts cultivated on slag and vitrified waste grew and proliferated, forming thick layers on the specimens after 14 days. After 2 days of incubation, fibroblasts on slag and vitrified waste showed philopodiae (P) contacting the material surface. This is a typical morphology for vital cells (Figure 3).

Semiquantitative Examinations Using the Agar Diffusion Overlay Test

The validation of the test was confirmed by negative (bottle glass, empty device) and positive (0.5 M NaOH) controls (Figure 4). According to the response indices, the effect on the soluble compounds of the untreated fly ash and the positive control was almost the same. The region of smaller-diameter cells indicated a less cytotoxic effect of treated fly ash, slag, and vitrified waste. As shown in Figure 4 the transition from dark to light, i.e., active to nonactive cells, was rather sharp.

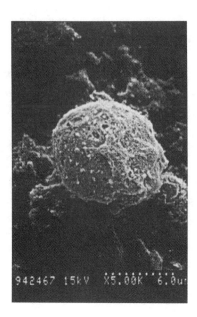

FIGURE 2. Fibroblast cells on fly ash after 2 days of incubation (left: 500x, right: 5,000x). The lack of philapodiae (see Figure 3, P), the presence of a porous cell membrane, and the spherical shape are indications of avital cells.

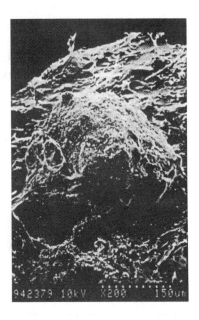
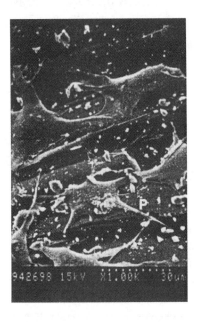

FIGURE 3. Fibroblast cells on slag (left, 200x) and vitrified waste (right, 1,000x) after 2 days of incubation. The development of philapodiae (P) and spread shape are indications of vital cells.

Material	Response Index (RI = ZI/LI)
negative control 1 (empty cylinder)	0 / 0
negative control 2 (bottle glass)	1 / 0
positive control (0.5 M NaOH)	5 / 5
FU	5 / 5
FU : bottle glass (1:1)	5 / 5
FU HCl	2 / 5
FU HNO₃	2 / 5
FU H₂SO₄	2 / 5

FIGURE 4. Agar diffusion overlay test (ASTM F895-84). The higher the zone and the lysis indices, the more toxic the test material. The photo shows the rather sharp transition from dark to light, i.e., from active to inactive cells.

Quantitative Colorimetric Measurements Using Direct Contact Methods

With increasing concentrations of sample added to the culture medium, the dehydrogenase activity of the fibroblasts decreased asymptotically to a concentration-independent value. According to these levels, the cytotoxicity of the examined samples (see Table 1 for abbreviations) decreases in the following order (Figure 5): FM, FU >> FM HCl, FU HCl, SUS > SWS, SWS HCl >> MHV, FAV, BGV, PSV.

Quantitative Colorimetric Measurements Using Indirect Methods

Quantitative measurement showed that slag (SUS) was less toxic than treated fly ash (FU HCl, FU H_2SO_4, FU HNO_3), which was again less toxic than untreated fly ash. This resulted from the dilution factors at optical density 0.5 ($OD_{0.5}$), which were about 3.2 for slag, 5.1 for treated fly ash, and 11 for untreated fly ash (Figure 6). The lower the $OD_{0.5}$, the less toxic the extracts.

DISCUSSION

In the case of the quantitative colorimetric measurements using direct contact methods (Figure 5), not only the soluble compounds, but also the sample surfaces had an effect on the cell activity. If the concentration of the sample is

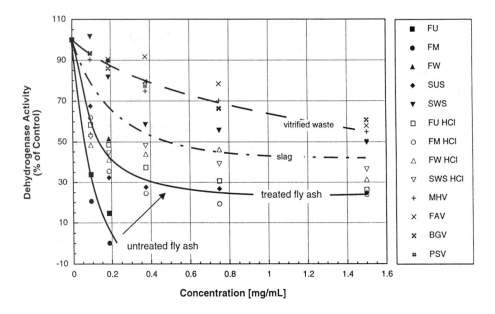

FIGURE 5. Quantitative colorimetric measurements using direct contact methods. The direct contact toxicity decreases in the order FM, FU >> FM HCl, FU HCl, SUS > SWS, SWS HCl >> MHV, FAV, BGV, PSV.

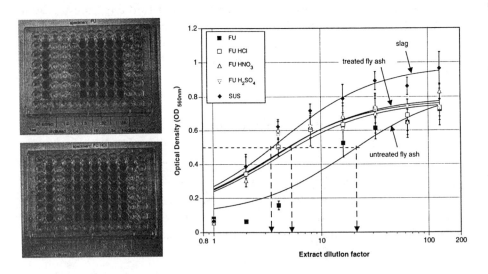

FIGURE 6. Quantitative colorimetric measurements using indirect contact methods. Additional treatment reduced the acute toxicity of fly ash. The dilution factors at 0.5 optical density (OD_{50}) are 3.2 for slag, 5.1 for treated fly ash, and 11 for untreated fly ash.

higher thant 0.4 mg/mL, the entire bottom of the culture dishes was covered with the sample material. Therefore, the cells' dehydrogenase activity quantified by the MTT test was dependent on the direct toxicity of the test specimen.

As shown above, cell cultures are very sensitive to any changes to the normal incubation conditions. Therefore, it is important that apparent changes are due to one single parameter. In-direct contact methods there are surface (topography, hydrophobicity) and chemical (soluble compounds) parameters to be considered, as well as pH changes, because the sample cannot be washed before being seeded with cells. Indirect methods, especially the extraction method, allows the pH to be adjusted, and there is no influence of the surfaces. Therefore, further investigations will concentrate on the extraction method where the extracts are obtained according to standard test methods used in waste management.

CONCLUSION

All analytical methods gave a consistent result: untreated fly ash was the most toxic, and vitrified samples the least toxic. Additional treatment of fly ash reduced its cytotoxicity practically to the level of slag. Therefore, cell culture techniques are considered to be a valid method for monitoring the acute toxicity of environmentally hazardous materials.

ACKNOWLEDGMENTS

The work presented is part of the Swiss National Science Foundation's Priority Programme "Environmental Technology and Environmental Research" (SPP U), which was launched in 1992. Data and materials were kindly provided by B. Dettwiler et al. (Sulzer Chemtech, Winterthur, Switzerland), P. Amann et al. (Institute of Geotechnology, Federal Institute of Technology, Switzerland), H. Vonmont et al. (Swiss Federal Laboratories for Materials Testing and Research, Dübendorf, Switzerland), and P. Kesselring et al. (Paul Scherrer Institute, Villigen, Switzerland), all of whom were participants of the SPP U program.

REFERENCES

Ahne, W. 1985. "Untersuchungen über die Verwendung von Fischzellkulturen für Toxizitätsbestimmungen zur Einschränkung und Ersatz des Fischtests." *Zentralblatt fuer Bakteriologie, Mikrobiologie und Hygiene, 1. Abt Orig B, 180*: 480-504.

ASTM F895-84. 1984. "Standard test method for agar diffusion cell culture screening for toxicity." American Society for Testing and Materials.

Babich, H., C. Shopsis, and E. Borenfreund. 1986. "In vitro cytotoxicity testing of aquotic pollutants (cadmium, copper, zinc, nickel) using established fish cell lines." *Ecotoxicology and*

Barrett, J., M.N. Hughes, G.I. Karavaiko, and P.A. Spencer. 1993. "Metal extraction by bacterial oxidation of minerals." In E.H.J. Burgess (Ed.), *Ellis Horwood Series in Inorganic Chemistry*, Ellis Horwood Limited, Chichester.

Burgstaller, W., and F. Schinner. 1993. "Leaching of metals with fungi." *Journal of Biotechnology* 27: 91-116.

Hansen, P.D. 1989. "Ein Fischzellkulturtest als Ergänzungs- oder Ersatzmethode zum Fischtest." *Bundesgesundheitsblatt* 32 (8): 317-321.

ISO 10993-5. 1992. *Evaluation of medical devices—Part 5:* "Tests for cytotoxicity: in vitro methods." International Standardization Organization .

Jellum, E., A. K. Thorsrud , N. C. Herud, H. J. Tong, and F. W. Karasek. 1992. "The Evaluation of Toxicity of Water Leachates from Incinerator Fly Ash Using Living Human Cells and Analytical Techniques." In R.E. Clement, K.W.M. Siu, and H.H. Hill Jr. (Eds.), *Instrumentation for Trace Organic Monitoring.* Lewis Publishers, Inc., Boca Raton, FL.

Lahmann, E., J. Knie, R. Kanne, and P.D. Hansen. 1989. "Biologische Toxizätstests." *Umweltwissenschaften und Schadstoff-Forschung, Zeitschrift für Umweltchemie und Ökotoxikologie 3:* 20-27.

Sewing, K.-F. (Ed.). 1994. *Zum Ersatz von Tierversuchen.* Kommentare-Projektberichte-Literaturbesprechungen. Ausgewählte Beiträge aus dem Informationsblatt In-vitro-Systeme 1986-1993. Schlütersche Verlagsanstalt und Druckerei GmbH & Co., Hannover, Germany.

The Biodegradation of Fluoranthene as Monitored Using Stable Carbon Isotopes

Beth A. Trust, James G. Mueller,
Richard B. Coffin, and Luis A. Cifuentes

ABSTRACT

The measurement of stable isotope ratios of carbon ($\delta^{13}C$ values) was investigated as a viable technique to monitor the intrinsic bioremediation of polycyclic aromatic hydrocarbons (PAHs). Biometer flask experiments were conducted in which the bacterium, *Sphingomonas paucimobilis*, designated EPA505, was grown on fluoranthene. During growth of EPA505 on fluoranthene, bacterial biomass, respired CO_2, and dissolved organic carbon (DOC), as well as fluoranthene, were sampled over 8 days. The concentrations and $\delta^{13}C$ values of each of these carbon pools were determined. The concentration of fluoranthene decreased from 12.1 ± 2.0 (n = 2) to 3.0 ± 0.9 (n = 2) mg C per flask over 188 h, and CO_2 increased from undetectable levels to 7.1 ± 0.3 (n = 4) mg C per flask. A total of 55.5% mineralization resulted. DOC concentrations remained fairly constant with time, averaging 2.2 to 3.6 mg C per flask. The $\delta^{13}C$ value of fluoranthene remained constant over the course of the experiment, averaging –24.5 ± 0.2 ‰ (n = 8). Bacterial nucleic acids and respired CO_2 took on $\delta^{13}C$ values similar to those of fluoranthene within 47 h, measuring –22.6 and –24.3 ‰, respectively.

INTRODUCTION

A challenge in the bioremediation of contaminated field sites is establishing that organisms are actually effecting the cleanup of the site. This is especially true for in situ bioremediation applications (U.S. EPA 1994). A novel approach for proving that bacteria and/or other organisms biodegrade pollutant compounds in situ is to compare the $\delta^{13}C$ values of the pollutants, bacteria (or a fraction thereof), and CO_2. For some time now, $\delta^{13}C$ values have been used in ecological research to study the flow of organic material through food webs (Rundel et al. 1988). In pollutant research, Aggarwal and Hinchee (1991) demonstrated that soil gas CO_2 at contaminated sites differs isotopically from that at uncontaminated sites. The goal of this study was to determine if bacteria and respired CO_2 have $\delta^{13}C$ values similar to carbon sources, thus establishing the flow of carbon from contaminant into bacterial biomass and then into respired CO_2. Documentation that $\delta^{13}C$ values are useful tracers of biodegradation in controlled laboratory experiments is prerequisite to applying this technique to the study of bioremediation at actual field sites.

EXPERIMENTAL PROCEDURES AND MATERIALS

Biometer Flask Experiment

Sphingomonas paucimobilis (EPA505) was grown on 1.8 g of glucose ($\delta^{13}C$ = –10.1 ‰) as the sole carbon source in 500 mL of minimal salts medium (Bushnell-Haas). Cultures were incubated at 30°C with vigorous mixing. CO_2 was collected into 3 mL of 2N NaOH contained in a test tube in the culture flask to measure the $\delta^{13}C$ value of CO_2 from cells grown on glucose. Bacteria were harvested after 48 h by centrifugation (10,000 rpm, 10 min, 4°C), resuspended and washed in phosphate buffer (50 mM, pH 7.2), and centrifuged again. The cell pellet was resuspended in 2 mL of phosphate buffer. Cell concentrations were determined by direct plate counts.

Biometer flasks were fitted with two stoppers: (1) a stopper above the 250-mL incubation flask containing an ascarite/drierite CO_2 trap that was closed to the sample by a stopcock, and (2) a stopper above the sidearm containing a 15-gauge needle through which NaOH samples could be withdrawn. Into the sidearm of each of 20 flasks was placed 2 to 3 mL of CO_2-free, 2N NaOH to collect respired CO_2. In the incubation section of each biometer flask, 60 mL of Bushnell-Haas medium was added to 13.5 mg of fluoranthene ($\delta^{13}C$ = –24.4 ‰) to produce a fluoranthene concentration of 225 mg/L. Flasks were sonicated for 30 seconds to break up large fluoranthene crystals. An aliquot of the EPA505 culture was added to 16 of the "live" biometer flasks to produce a cell concentration of 10^8 CFU/mL. An identical amount of autoclaved cells was added to each of the remaining four killed-cell control flasks. Four "live" flasks were retained for initial (t = 0) sampling, and the remaining flasks were shaken gently (120 rpm) in the dark at a temperature of 30°C for 47, 96, or 188 h. The four killed-cell control flasks were incubated under the same conditions for 188 h.

At each sampling point, four "live" flasks were sacrificed in their entirety. The NaOH containing respired CO_2 was withdrawn from the sidearms of each of the flasks and injected into crimp-sealed vials that had been flushed with N_2 gas. Two flasks were then sampled for fluoranthene concentration and $\delta^{13}C$ values by extraction of the entire flask contents with methylene chloride. The other two flasks were sampled for DOC and bacteria. The contents of these flasks were pipetted into FEP (fluorinated ethylene propylene) centrifuge tubes and centrifuged at 10,000 rpm for 10 min. The supernatant was filtered through glass fiber filters (GF/F) for determination of DOC concentration on a total organic carbon (TOC) analyzer, and the bacterial pellet was frozen at –70°C for nucleic acid extraction. At 188 h, the four killed-cell control flasks were sampled along with four "live" flasks for direct comparison of biotic and abiotic changes in measured parameters. Nucleic acids were not sampled in the control flasks because the concentrations of cells were not believed to be large enough for an evaluation of $\delta^{13}C$ to be made.

Analyses of Concentrations

DOC concentrations were measured using a Shimadzu 5000 TOC analyzer (high temperature, platinum catalyst). Fluoranthene and CO_2 concentrations were determined by gas chromatography (GC). For fluoranthene, the methylene chloride extracts were injected onto a Hewlett-Packard (HP) 5890 Series II/3 GC equipped with a 25-m × 0.32-mm HP-5 column and a flame ionization detector (FID). CO_2 concentrations were measured by withdrawing a 25- to 500-μL subsample of the NaOH and injecting it into a crimp-sealed headspace vial containing 250 μL of concentrated phosphoric acid. The liberated CO_2 was analyzed on an HP 5890 Series II/3 gas chromatograph equipped with an automatic headspace analyzer and a 20-m × 0.53-mm Poraplot Q fused silica column. CO_2 was reduced in the presence of a Ni catalyst to CH_4, and CH_4 concentrations were measured with the FID.

Stable Carbon Isotope Analyses

All carbon isotope ratios ($^{13}C/^{12}C$) are expressed by the conventional "delta" (δ) notation:

$$\delta^{13}C \ (‰) = [(R_{sample}/R_{std}) - 1] \times 1,000 \tag{1}$$

where R_{sample} and R_{std} are the $^{13}C/^{12}C$ isotope ratios of the sample and standard, respectively (Craig 1957). All values are reported relative to the conventional Peedee belemnite standard. The working standard for our carbon isotope measurements was tank CO_2 ($\delta^{13}C_{PDB} = -5.5 ‰$).

The $\delta^{13}C$ values of glucose and fluoranthene crystals were measured using conventional methods (Macko 1981), as were the $\delta^{13}C$ values of nucleic acids, which were extracted from the bacteria using a phenol:chloroform extraction method (see Coffin et al. 1990 for details). Samples were placed into precombusted quartz tubes. Nucleic acid samples, which were dissolved in distilled water, were then freeze-dried for 24 h. Copper wire (5 g) and CuO (2.5 g) were added to the samples, and the tubes were evacuated and sealed. In turn, the tubes were heated to 850°C for 2 h to generate CO_2 gas, which was isolated cryogenically on a vacuum line. The carbon isotope content of the CO_2 was measured on a Finnigan MAT 252 isotope ratio mass spectrometer having an overall precision of ±0.1 ‰.

In contrast to nucleic acids and crystalline samples, the $\delta^{13}C$ values of the fluoranthene samples extracted from the biometer flasks and of respired CO_2 were measured using a novel GC/ITMS/IRMS system. This system consists of three parts: (1) a Varian Star 3400 gas chromatograph (GC), (2) a Magnum (Finnigan) ion trap mass spectrometer (ITMS), and (3) a Delta S (Finnigan) isotope ratio mass spectrometer (IRMS). With this system, mixtures of analytes are separated by the GC column. Of the total effluent, 10% is carried into the ion trap mass spectrometer for compound identification. The remaining effluent passes through a 940°C combustion furnace packed with cupric oxide where carbon constituents are reacted to CO_2 gas. The CO_2 then flows into the isotope ratio mass spectrometer, where its carbon isotope composition is measured. Reproducibility of $\delta^{13}C$ values for multiple injections of fluoranthene and CO_2 standards by this instrument was better than ±0.2 ‰. For the determination of the $\delta^{13}C$ values of fluoranthene, 1 μL of the methylene chloride extracts was injected into the GC, which was equipped with a 50-m × 0.32-mm i.d. Ultra-2 (Hewlett-Packard) column. For the determination of the $\delta^{13}C$ values of respired CO_2, an aliquot of the NaOH sample was injected into a headspace vial containing 250 μL of concentrated phosphoric acid. From 50 to 500 μL of the headspace was injected onto the GC/ITMS/IRMS system. The GC was equipped with a 25-m × 0.32-mm Poraplot Q column.

RESULTS

Concentrations (mg C per flask) of fluoranthene, DOC, CO_2 and total combined carbon concentrations at various sampling times throughout the experiment are reported in Table 1. Initially 12.8 mg C of fluoranthene (13.5 mg FLA) was added to the flasks. At t = 0, the measured amount of fluoranthene averaged 12.1 ± 2.0 mg C per flask (n = 2) , indicating a recovery efficiency of 94.5%. By 47 h, the concentration of fluoranthene was reduced to 8.1 ± 1.1 mg C per flask (n = 2), and it continued decreasing until a final value of 3.0 ± 0.9 mg C per flask (n = 2) was reached at 188 h. Killed-cell control flasks had an average fluoranthene concentration of 9.2 ± 0.8 mg C per flask after 188 h, indicating that approximately 29% abiotic loss occurred.

In contrast to fluoranthene, DOC concentrations remained fairly constant throughout the experiment. They ranged between 2.2 and 3.6 mg C per flask. Killed-cell controls averaged 3.9 ± 0.3 mg C per flask (n = 2), similar to the values for all "live" flasks.

Under active growth conditions, CO_2 concentrations increased from below detection to 7.1 ± 0.3 mg C per flask (n = 4). Percent mineralization values were calculated based on an initial fluoranthene concentration of 12.8 mg C per flask. After 188 h, 55.5% mineralization of fluoranthene to CO_2 occurred in the "live" flasks, whereas only 6.3% mineralization occurred in the "control" flasks.

Total carbon concentrations were calculated by summing the concentrations of fluoranthene, DOC, and CO_2 at each time point. They averaged 14.1 ± 3.8 mg C per flask (n = 20). As stated earlier, 12.8 mg C were added to each

TABLE 1. Concentrations of fluoranthene (FLA), dissolved organic carbon (DOC), and CO_2 at various times throughout the experiment.[a]

| Time (h) | mg C per flask ± 1 s.d. (n) | | | | % Mineralization |
	FLA	DOC	CO_2	Total	
0	12.1 ±2.0 (2)	2.2 ±0.1 (2)	0.0 ±0.4 (4)	14.3 ±2.0	0.0
47	8.1 ±1.1 (2)	2.9 ±0.2 (2)	4.1 ±0.7 (4)	15.1 ±1.3	32.0
96	5.2 ±1.3 (2)	3.6 ±0.1 (2)	5.2 ±0.8 (4)	14.0 ±1.5	40.6
188	3.0 ±0.9 (2)	3.3 ±0.3 (2)	7.1 ±0.3 (4)	13.4 ±0.9	55.5
KCC 188	9.2 ±0.8 (2)	3.9 ±0.3 (2)	0.8 ±2.2 (4)	13.9 ±2.4	6.3

(a) Values are reported as mg C per flask. The initial amount of fluoranthene added to each flask was 12.8 mg C. Percent mineralization = 100 × (amount of carbon liberated as CO_2 ÷ 12.8). KCC 188 = Killed-cell control at 188 h.

flask as fluoranthene. The only other source of carbon was bacteria. Assuming an average carbon content of 30 fg C per cell (Simon & Azam 1989) at an initial cell concentration of 10^8 CFU/mL in 60 mL, 0.2 mg C was added initially as bacterial biomass. The combination of fluoranthene and bacteria produces a total of about 13 mg C, slightly lower but within 92% of the average value of 14.1 mg C calculated for the flasks.

The measured $\delta^{13}C$ values for fluoranthene, bacterial nucleic acids, and respired CO_2 are illustrated in Figure 1. The $\delta^{13}C$ value of fluoranthene remained constant with time, averaging –24.5 ± 0.2 ‰ (n = 8). This value, determined with the GC/ITMS/IRMS system, agrees well with the value of –24.4 ‰, which was measured for fluoranthene using more conventional techniques (Macko 1981). A value of –24.6 ± 0.1 ‰ (n = 2) was measured for fluoranthene isolated from the killed-cell control flasks.

Bacterial nucleic acids initially had a $\delta^{13}C$ value of –13.6 ‰, close to but slightly more negative than the value of glucose ($\delta^{13}C$ = –10.1 ‰). Within 47 h bacterial nucleic acids reached and maintained a $\delta^{13}C$ value closer to that of fluoranthene, averaging –22.7 ± 0.3 ‰ (n = 6). This value is 1.8 ‰ more positive than that of fluoranthene. Other studies have indicated that bacterial nucleic acids tend to be 2 ‰ enriched in ^{13}C relative to their carbon source (Coffin et al. 1989; Coffin et al. 1990); however, there were a few cases cited in

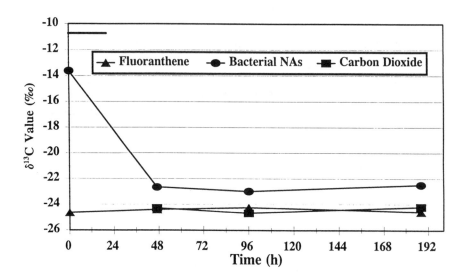

FIGURE 1. $\delta^{13}C$ values of fluoranthene, bacterial nucleic acids (NAs), and respired carbon dioxide over the course of the experiment. The short line at $\delta^{13}C = -10.8$ ‰ indicates the $\delta^{13}C$ value of respired CO_2 collected from the EPA505 culture as it was grown on glucose ($\delta^{13}C = -10.1$ ‰). The $\delta^{13}C$ values of fluoranthene and respired CO_2 in the killed-cell control flasks averaged –24.6 ‰ and –23.0 ‰, respectively, at 188 h.

those same studies in which nucleic acids were found to be depleted in ^{13}C relative to the carbon source. It is apparent that further investigations into the $\delta^{13}C$ values of nucleic acids isolated from bacteria growing on sole carbon sources need to be conducted. It is quite possible that the isotopic fractionation factor between carbon source and bacterial nucleic acids is both species specific and source specific.

Similar to nucleic acids, respired CO_2 quickly took on the $\delta^{13}C$ value of fluoranthene. Within 47 h of feeding EPA505 fluoranthene, the $\delta^{13}C$ value of CO_2 averaged -24.3 ± 0.6 ‰ (n = 4). This value differs significantly from the value of -10.8 ± 0.1‰ (n = 2) measured for CO_2 collected from EPA505 growing on glucose. A value of -23.0 ± 1.0 ‰ (n = 4) was measured for CO_2 collected from killed-cell control flasks at 188 h. This indicates that the CO_2 produced in these flasks resulted from the abiotic degradation of fluoranthene.

DISCUSSION

This study demonstrates that bacteria and respired CO_2 have $\delta^{13}C$ values that are very similar to that of their carbon source. This provides evidence that $\delta^{13}C$ values are valuable indicators of the source of carbon to microorganisms. The use of stable carbon isotope ratios would be useful in cometabolism studies in which microorganisms are fed a combination of two substrates having diverse $\delta^{13}C$ values. By measuring the $\delta^{13}C$ values of the nucleic acids of the organisms and their respired CO_2, it should be possible to determine if first one substrate is degraded and then the other, or if the two substrates are cometabolized.

In field settings, the success of in situ bioremediation of contaminated sites is usually based on a decrease in contaminant concentrations. When contaminant levels decrease, however, it is difficult to prove that the contaminants actually have been degraded or just have partitioned into an immobile phase or been transported to a new location. By measuring the $\delta^{13}C$ values of respired CO_2, bacterial biomass, contaminants, and indigenous carbon sources, in addition to contaminant concentrations, it should be possible to prove with a higher degree of certainty that biodegradation of contaminants is occurring.

REFERENCES

Aggarwal, P. K., and R. E. Hinchee. 1991. "Monitoring in situ biodegradation of hydrocarbons by using stable carbon isotopes." *Environmental Science and Technology* 25:1178-1180.

Coffin, R. B., B. Fry, B. J. Peterson, and R. T. Wright. 1989. "Carbon isotopic compositions of estuarine bacteria." *Limnology and Oceanography* 34(7):3205-3230.

Coffin, R. B., D. J. Velinsky, R. Devereux, W. A. Price, and L. A. Cifuentes. 1990. "Stable carbon isotope analysis of nucleic acids to trace sources of dissolved substrates used by estuarine bacteria." *Applied and Environmental Microbiology* 56(7):2012-2020.

Craig, H. 1957. "Isotopic standards for carbon and oxygen and correction factors for mass-spectrometric analysis of carbon dioxide." *Geochimica et Cosmochimica Acta* 12:133-149.

Macko, S. A. 1981. "Stable nitrogen isotope ratios as tracers of organic geochemical processes." PhD Dissertation, The University of Texas at Austin, TX.

Rundel, P. W., J. R. Ehleringer, and K. A. Nagy (Eds.). 1988. *Stable Isotopes in Ecological Research.* Springer-Verlag, New York.

Simon, M., and F. Azam. 1989. "Protein content and protein synthesis rates of planktonic marine bacteria." *Marine Ecology Progress Series* 51:201-213.

U.S. Environmental Protection Agency. 1994. "Up-to-date ways to assess when bioremediation works." *Ground Water Currents.* EPA-542-N-94-007.

Stable Carbon Isotope Analysis to Verify Bioremediation and Bioattenuation

Kevin D. Van de Velde, Michael C. Marley,
James Studer, and David M. Wagner

ABSTRACT

Carbon is composed primarily of two stable isotopes, carbon-12 and carbon-13. The ratio of carbon-12 to carbon-13 is approximately 99:1. Some variation in the stable carbon isotope ratio occurs due to fractionation during physical, chemical, and biological processes. It is this variation that allows evaluation of biological systems where carbon compounds are mineralized, resulting in the liberation of carbon dioxide. Using stable isotope analysis, it may be possible to determine the source(s) of the liberated carbon dioxide, thereby confirming successful mineralization of the targeted carbon compound(s). Currently, this type of determination is completed by a mass balance calculation, which is often difficult and expensive to employ under field conditions. Due to the relatively low cost of stable isotope analysis and the reduced variability in data from vapor sampling compared to soil sampling, stable isotope analysis can be an inexpensive and effective tool for monitoring and evaluating biological systems. This paper reviews stable isotope theory and recommended sampling procedures. Two case studies are presented that illustrate the application of stable isotope analysis for evaluating bioremediation systems and demonstrating bioattenuation. Additional potential applications of the stable isotope method for bioremediation evaluation and monitoring are discussed.

INTRODUCTION

Over the past few years, in situ bioremediation systems, such as bioventing and biosparging, have been applied to numerous sites to degrade petroleum hydrocarbons. The systems are designed to accelerate in situ biological degradation by enhancing the existing subsurface environment, stimulating indigenous aerobic microbes to metabolize the petroleum hydrocarbons.

Several researchers have shown that aerobic microbes collected from hydrocarbon release sites are often capable of completely mineralizing many hydrocarbons to carbon dioxide and water under laboratory conditions.

Several approaches have been used to provide evidence of biological degradation under field conditions. These approaches include measurement of changes in the concentration of hydrocarbons, temperature, number of hydrocarbon-degrading microorganisms, ratio of fast-degrading hydrocarbons to slowly degrading hydrocarbons, metabolic intermediates, oxygen uptake, and carbon dioxide generation (Aggarwal and Hinchee 1991). Of these approaches, only detailed mass balance calculations using measurements of the changes in the concentration of hydrocarbons and metabolic intermediates can provide direct evidence of biological degradation of the hydrocarbons.

Due to the nonhomogeneous distribution of hydrocarbons and their metabolic intermediates in most subsurface environments and the multiple sources and/or sinks that must be accounted for, it is generally impractical to perform a mass balance to verify biological degradation. With the significant growth in the use of bioremediation, new methods that are less expensive and more reliable must be developed to verify biological degradation in the field.

One method that has received some attention in the past as a possible alternative for verifying biological degradation of organic carbon compounds at hazardous waste sites is the use of stable carbon isotopes (Aggarwal and Hinchee 1991; Suchomel et al. 1990). This method involves the examination of carbon-containing respiration products from biodegradation to determine the source of the carbon. The carbon-containing end product from aerobic biodegradation of hydrocarbon typically is carbon dioxide.

Although changes in soil gas carbon dioxide concentrations have been used in the past as a means of demonstrating biodegradation, several other sources of carbon dioxide exist. These sources of carbon dioxide include the atmosphere, biological degradation of organic carbon compounds other than the hydrocarbon of interest, plant root respiration, and calcite precipitation. Carbon dioxide from sources such as these would need to be discerned. Through the use of stable carbon isotope analysis, the amount of carbon dioxide contributed from multiple sources, including carbon dioxide resulting from the biodegradation of the hydrocarbon of interest, can be determined.

THEORETICAL BACKGROUND

Carbon is comprised primarily of two stable isotopes, carbon-12 and carbon-13, in a ratio of approximately 98.89% carbon-12 to 1.11% carbon-13. The stable carbon isotope method is employed by comparing the stable carbon isotope composition of the carbon dioxide with the potential source material. Due to differences such as mass and nuclear spin of the carbon-12 and carbon-13 atoms, the carbon isotopes behave slightly differently when undergoing reactions (Galimov 1985). As a result, the isotopic composition of the product of a reaction may differ from the isotopic composition of the original reactants.

The change in isotopic composition of the reaction products relative to the reactants is referred to as fractionation and must be accounted for in the comparison of the end product with possible source substrates. If the fractionation associated with the biological degradation process can be estimated, then the isotopic composition of the original carbon substrate can be deduced.

The isotopic compositions of carbon-containing compounds can be measured by mass spectrometric analysis using a dual-inlet gas isotope ratio mass spectrometer (GIRMS). The isotopic composition by convention is referenced to the Peedee belemnite (PDB) standard and is reported as the $\delta^{13}C$ of the sample in parts per thousand (per mil, ‰) deviation from the PDB standard. The PDB standard, a Cretaceous calcite fossil (*Belemnitella americana*) is an arbitrary standard (Craig 1957). The $\delta^{13}C$ of the sample represents the relative difference between isotopic composition of the carbon sample (R_x) and of the PDB standard (R_s). The value of the $\delta^{13}C$ of the sample is determined as follows:

$$\delta^{13}C = \frac{R_X - R_S}{R_S} \times 1{,}000 \qquad (1)$$

A negative $\delta^{13}C$ value indicates that the sample is depleted in carbon-13 relative to the PDB standard.

For most natural carbon sources (atmospheric carbon dioxide, marine carbonate, nonmethane hydrocarbons, etc.), the $\delta^{13}C$ range is relatively narrow. For example, nonmethane petroleum hydrocarbons have been observed to range from –34 ‰ to –20 ‰ (Fuex 1977). Atmospheric carbon dioxide has been found to range between –12 ‰ to –7.4 ‰, depending on the level of air pollution in the area sampled (Coleman et al. 1991). Most forests have soil carbon with carbon isotopic values of –29 ‰ to –24 ‰ whereas tropical grasslands and savannahs have carbon isotopic values of –13 ‰ to –12 ‰ (Schimel 1993). Methane has been observed to possess the largest range in isotopic composition of any naturally occurring material, ranging from –110 ‰ to –13 ‰ (Fuex 1977; Whiticar et al. 1986). The wide variation in the $\delta^{13}C$ value of organic and inorganic carbon in the biosphere is due to fractionation during formation.

The fractionation of methane during formation (Cloud et al. 1958; Rosenfeld and Silverman 1959; Schoell 1980; Whiticar et al. 1986) and oxidation (Barker and Fritz 1981; Fuex 1977; Schoell 1980; Whiticar and Faber 1985; Zyakun et al. 1983) has been widely studied. The $\delta^{13}C$ of methane is greatly affected by its source and the fractionation associated with the formation process. Methane resulting from bacterial sources has been observed to have $\delta^{13}C$ values ranging from –90 ‰ to –50 ‰ (Fuex 1977), whereas methane from abiotic sources has $\delta^{13}C$ values heavier (i.e., more positive $\delta^{13}C$) than approximately –58 ‰ (Coleman et al. 1977). Caution should be used when applying these ranges to determine methane sources because biogenic methane, which has been subjected to partial oxidation by bacteria, can be fractionated so that the residual methane has a $\delta^{13}C$ in the range usually attributed to abiotic sources (Barker and Fritz 1981; Coleman et al. 1981). Coleman et al. (1981) suggested the additional use of stable hydrogen isotopes to assist in the identification.

Two distinct pathways of methanogenesis occur in natural environments: hydrogenation of carbon dioxide, and acetate fermentation (Chapelle 1993). Hydrogenation of carbon dioxide uses carbon dioxide in the production of methane, and acetate fermentation uses acetate in the production of methane and carbon dioxide. Methane production in landfills by these processes has resulted in residual (hydrogenation of carbon dioxide) and produced (acetate fermentation) carbon dioxide with $\delta^{13}C$ values ranging from +10.3 ‰ to as high as +20.0 ‰ relative to the PDB standard (Games and Hayes 1976; Baedecker and Back 1979).

Consider a hypothetical case in which two carbon-containing compounds are present in a single container, each of which undergoes a biological reaction in the presence of air to produce carbon dioxide. The $\delta^{13}C$ value of the total carbon dioxide present in the container can be related to the $\delta^{13}C$ values of the carbon dioxide from the two carbon-containing compounds by:

$$\delta^{13}C_{total} = \delta^{13}C_{atmos}\left\{\frac{[CO_2]_{atmos}}{[CO_2]_{total}}\right\} + \delta^{13}C_1\left\{\frac{[CO_2]_1}{[CO_2]_{total}}\right\} + \delta^{13}C_2\left\{\frac{[CO_2]_2}{[CO_2]_{total}}\right\} \qquad (2)$$

where: $\delta^{13}C_{total}$ = $\delta^{13}C$ of the total resultant carbon dioxide from the reactions

   $\delta^{13}C_{atmos}$ = $\delta^{13}C$ of the atmospheric carbon dioxide

   $\delta^{13}C_1$ = $\delta^{13}C$ of the carbon dioxide resulting from the reaction of the first carbon-containing compound

   $\delta^{13}C_2$ = $\delta^{13}C$ of the carbon dioxide resulting from the reaction of the second carbon-containing compound

   $[CO_2]_{total}$ = total moles of carbon dioxide

   $[CO_2]_{atmos}$ = moles of atmospheric carbon dioxide present in container

   $[CO_2]_1$ = moles of carbon dioxide resulting from the reaction of the first carbon-containing compound

   $[CO_2]_2$ = moles of carbon dioxide resulting from the reaction of the second carbon-containing compound.

This equation assumes that no other sources and/or sinks of carbon dioxide are present in the container.

By mass balance, the total moles of carbon dioxide present in the container are equal to the sum of the moles of atmospheric carbon dioxide and carbon dioxide resulting from the reactions of the first and second carbon-containing compounds. Inserting the mass balance, simplifying, and solving for $[CO_2]_2$ yields:

$$[CO_2]_2 = \frac{[CO_2]_{total}\left(\delta^{13}C_{total} - \delta^{13}C_1\right) + [CO_2]_{atmos}\left(\delta^{13}C_1 - \delta^{13}C_{atmos}\right)}{\left(\delta^{13}C_2 - \delta^{13}C_1\right)} \quad (3)$$

If the atmospheric carbon dioxide contribution can be considered insignificant (i.e. the number of moles of atmospheric carbon dioxide, $[CO_2]_{atmos}$ contributed is small compared to the total number of moles of carbon dioxide resulting from the reaction of the carbon-containing compounds), Equation 3 is reduced to:

$$[CO_2]_2 = [CO_2]_{total}\left\{\frac{\left[\delta^{13}C_{total} - \delta^{13}C_1\right]}{\left[\delta^{13}C_2 - \delta^{13}C_1\right]}\right\} \quad (4)$$

Equation 4 is essentially the same equation used by Fry et al. (1978) in studies of grasshopper diets using stable carbon isotope analysis. Because $[CO_2]_{total}$ and $\delta^{13}C_{total}$ are easily measured, it can be seen that $[CO_2]_2$ could be solved for by Equation 4 if $\delta^{13}C_1$ and $\delta^{13}C_2$ were known. Once $[CO_2]_2$ is known, $[CO_2]_1$ could be calculated. The $\delta^{13}C_1$ and $\delta^{13}C_2$ are related to the measurable $\delta^{13}C$ of the original carbon-containing compounds by accounting for the fractionation that the carbon from the two substrates experienced during the reactions. This fractionation may be described by (Parker and Calder 1970):

$$\alpha = \delta^{13}C_P \Big/ \delta^{13}C_S \quad (5)$$

where, α = fractionation coefficient
   $\delta^{13}C_S$ = $\delta^{13}C$ of the original carbon substrate
   $\delta^{13}C_P$ = $\delta^{13}C$ of the produced carbon dioxide ($\delta^{13}C_1$, $\delta^{13}C_2$)

Equation 5 assumes that the reservoir of source carbon that the microorganisms oxidize is large compared to the amount of source carbon oxidized. Given a finite source of carbon substrate, initially the substrate may become enriched in carbon-13 due to fractionation during the oxidation process. Conversely, the reaction end products will be depleted in carbon-13. However, as the source carbon becomes enriched in the carbon-13, the carbon-13 content of the end product of the reaction will increase. If the reaction goes to completion, the composition of the end products will approximate that of the source carbon (Schimel 1993).

As can be seen by the typical $\delta^{13}C$ ranges for different carbon sources given earlier, some overlap in the typical $\delta^{13}C$ ranges between different carbon sources is present. Should the $\delta^{13}C$ values of the carbon sources be close, discrimination between the carbon dioxide resulting from their oxidation may be difficult. In some cases, discerning the source of carbon dioxide resulting from potential sources with similar $\delta^{13}C$ values can be accomplished by considering the fractionation occurring during the oxidation processes. For instance, suppose two potential sources with similar $\delta^{13}C$ values are present. In addition, the oxidation process of the first source results in carbon dioxide that has a $\delta^{13}C$ value heavier than the substrate, and oxidation of the second source results in carbon dioxide that is lighter than the substrate. If the difference between the expected $\delta^{13}C$ values of the carbon dioxide resulting from the two oxidation reactions is sufficient, then the source of carbon dioxide resulting from the oxidation of one or both of the compounds may be determined.

Methane oxidation under aerobic conditions by methanotrophs has been observed to highly fractionate carbon. Carbon dioxide resulting from the oxidation of methane has been observed to be from 5.0 ‰ to 29.6 ‰ lighter than that of the methane substrate (Barker and Fritz 1981).

To date, few data are available regarding fractionation of nonmethane organic compounds undergoing aerobic biodegradation. In general, it has been observed that the $\delta^{13}C$ of carbon dioxide present in soil not affected by xenobiotic organic compounds is similar to that of the natural organic carbon present in the soil (Aggarwal and Hinchee 1991; McMahon et al. 1990), indicating that little fractionation occurs during the aerobic oxidation of natural organics. Aggarwal and Hinchee (1991) reported that carbon dioxide from soil gas in an uncontaminated area of Tyndall Air Force Base had an average $\delta^{13}C$ value of −18.4 ‰. The natural organic matter present in the soil had an average $\delta^{13}C$ value of −22.2 ‰. Accounting for the presence of atmospheric carbon dioxide, assuming 0.03% carbon dioxide with an average $\delta^{13}C$ value of −7.4 ‰, by Equation 4, the carbon dioxide is approximately 3.1 ‰ heavier than that of the source natural organics in the soil.

Aggarwal and Hinchee (1991) observed that carbon dioxide apparently resulting from the in situ aerobic biodegradation of petroleum hydrocarbons had $\delta^{13}C$ values approximately 3 ‰ heavier than that of the source hydrocarbons. This fractionation of 3 ‰ assumes that the carbon dioxide results exclusively from the oxidation of the petroleum hydrocarbons and not a mixture with carbon dioxide from the natural organics. Suchomel et al. (1990) analyzed the carbon dioxide present in the soil, where it was believed that trichloroethylene (TCE) was being aerobically biodegraded. They found that the $\delta^{13}C$ values of the carbon dioxide were within a range consistent with a hydrocarbon-derived source material. The authors are currently conducting funded research into the observed fractionation of stable carbon isotopes during the aerobic mineralization of mixed chlorinated and nonchlorinated hydrocarbons by microorganisms. This research may allow a more refined evaluation of the fractionation of organic compounds undergoing aerobic biodegradation.

CASE STUDY #1

Site Description

The site in question is an active fire training facility located in the northeast. Past fire training practices included the ignition of No. 2 fuel oil at several locations throughout the site. Upon extinguishing the training fires, the fire fighting fluids and cleanup wash water, containing dissolved and separate phase fuel oil, flowed into nearby unlined dry wells. Consequently, the groundwater and subsurface soils below the site were contaminated with high concentrations of petroleum hydrocarbons in three areas, as shown on Figure 1.

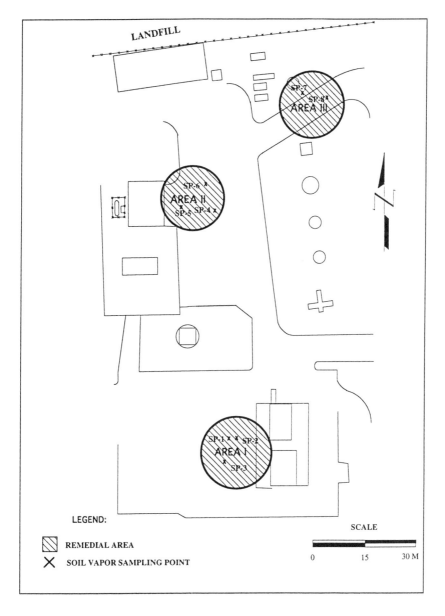

FIGURE 1. Areas of contamination for Case Study #1.

Due to the concentrations of hydrocarbons in the vadose zone soils and the probable biodegradability of the hydrocarbons, bioventing was selected for further study as part of the remedial approach. Bioventing is the process of stimulating indigenous microorganisms, generally through aeration of the vadose zone soils, to biologically degrade contaminants into carbon dioxide and water. The effectiveness of bioventing, as a remedial approach for the site, was evaluated during field pilot testing conducted at the three areas from December 1993 through April 1994.

Background soil gas measurements, obtained in each area prior to initiation of the pilot testing, indicated depressed levels of oxygen and elevated levels of carbon dioxide. These measurements indicated that biological activity had occurred or was occurring in the subsurface environment. High levels of methane, ranging from 5 to 10%, were present in each of the three areas prior to bioventing. The source of the subsurface methane was not known. However, it was postulated that the source of the methane was (1) the adjacent landfill, and/or (2) anaerobic production of methane in anoxic zones in the subsurface below the site.

The bioventing field pilot testing, conducted in all three areas, indicated positive biological trends of oxygen depletion and carbon dioxide evolution, as compared to background soil gas data. However, because both No. 2 fuel oil and methane can both be metabolized under aerobic conditions, the source of the carbon dioxide was undefined. A determination of the source of the carbon dioxide was therefore required to verify that the targeted substrate, No. 2 fuel oil, was being effectively biodegraded as a result of the bioventing system.

Methodology

Carbon dioxide could result from aerobic biodegradation of fuel oil and/or methane; therefore, stable carbon isotope analysis was employed to determine the source(s). Carbon dioxide samples were collected from three different sample points each in Area I and Area II. Two different sample points were sampled in Area III. Two methane samples were collected in Area III from the sample points used to collect the carbon dioxide samples. Further, a sample of the No. 2 fuel oil present in Area I was collected.

Measurements of the carbon dioxide, volatile hydrocarbon, and methane concentrations in the soil vapor at the time of sample collection were made prior to the sampling. The soil vapor across the site had carbon dioxide concentrations ranging from 6% to 12%. Volatile hydrocarbon (primarily toluene) was present in low concentrations at Area I during and immediately after the bioventing activity. Volatile hydrocarbons were not detected in the soil vapor in Areas II and III during the bioventing activity. Methane was not detected in the soil vapor of the three areas during and immediately after the bioventing activity.

Vapor samples were collected with a vacuum pump, which discharged into a 1-L Tedlar™ film bag with polypropylene valves. When collecting samples for carbon isotope analysis, at least 50 μg of carbon (carbon dioxide, methane, gasoline vapors, or liquid, etc.) will be necessary to complete the analysis using a GIRMS. More than 200 μg of carbon is preferred for completion of carbon isotope analysis.

For gas samples, approximately 100 mL of gas at standard temperature and pressure (STP), containing at least 2% of the carbon to be analyzed, is preferred. The gas samples can be placed into evacuated tubes, although Tedlar™, Teflon™, or Mylar™ bags with either stainless steel or polypropylene fittings are preferred. The construction material of the bags should be selected based on its compatibility with the sample gas. Sample bags simplify the extraction and handling of the gas sample in the lab, because the gas sample is at atmospheric pressure. Sample bags are commercially available with a septum built into the valve that allows insertion of a needle to extract the gas sample from the bag. The bags should be placed in a container that will prevent damage to the inflated bags during shipping. It will be important to notify the laboratory of the various gases (i.e., air, carbon dioxide, methane, gasoline-range petroleum hydrocarbons, etc.) to expect in the vapor samples. The types of gases present in the vapor sample will dictate the separation procedure used in the laboratory.

Volatile liquid product samples should be collected in a glass vial with a Teflon™ septum. Liquid samples should be maintained at 4°C during handling and shipping to prevent degradation of the sample.

Based on the literature previously cited, if carbon dioxide was the result of the aerobic degradation of nonmethane hydrocarbons or natural organic, it could be expected that the $\delta^{13}C$ of the carbon dioxide would be within approximately 3 ‰ heavier than the nonmethane carbon substrate. Carbon dioxide resulting from the oxidation of methane could be expected to have a $\delta^{13}C$ from 5.0 to 29.6 ‰ lighter than the methane substrate. Intermediate values would indicate a combination of sources for the total carbon dioxide present in the subsurface.

Due to the distribution of petroleum hydrocarbons across the site, background carbon dioxide samples could not be collected to evaluate the $\delta^{13}C$ of carbon dioxide resulting from the oxidation of nonhydrocarbon natural organics present at the site. However, because the natural carbon content of the vadose zone soils (fine to medium grained quartz sand) at the site is likely low, the contribution of carbon dioxide from oxidized natural organics was expected to be minimal.

Results and Discussion

The $\delta^{13}C$ values determined by GIRMS of the carbon dioxide, methane, and No. 2 fuel oil samples are shown on Table 1. The results of the carbon isotope analysis indicate that the source of the carbon dioxide present in the subsurface in Area I and Area II is likely No. 2 fuel oil. This is evidenced by the close agreement of the carbon dioxide $\delta^{13}C$ with the $\delta^{13}C$ of the No. 2 fuel oil substrate. Based on the literature previously cited, had the carbon dioxide resulted exclusively from the oxidation of methane, the $\delta^{13}C$ value of the

TABLE 1. Results of carbon dioxide concentration readings and stable carbon isotope analysis, with all $\delta^{13}C$ reported relative to the PDB standard.

Carbon Dioxide Samples		
Sample No.—Location	Carbon Dioxide, %	$\delta^{13}C$ of Carbon Dioxide, ‰
Sample Point #1—Area I	12	−27.3
Sample Point #2—Area I	10	−25.8
Sample Point #3—Area I	7	−25.9
Sample Point #4—Area II	8	−26.3
Sample Point #5—Area II	6	−26.7
Sample Point #6—Area II	6	−25.9
Sample Point #7—Area III	11.5	−20.0
Sample Point #8—Area III	7.5	−20.1
Methane and No. 2 Fuel Oil Samples		
Sample No.—Location	Carbon Source	$\delta^{13}C$ of Carbon Source, ‰
Sample Point #1—Area I	No. 2 Fuel Oil	−26.9
Sample Point #8—Area III	Methane	−50.3
Sample Point #8—Area III	Methane	−52.3

carbon dioxide would be expected to range from −55.3 ‰ to −81.9 ‰. The stated conclusion for the source of the carbon dioxide is further supported by:

1. Nondetection of methane in the soil gas in Areas I and II during and immediately after the bioventing operation, which indicates that methane may not have been present in these areas at concentrations sufficient to produce the carbon dioxide.
2. The presence of greater than 10^{-5} contaminant-specific (fuel oil) colony-forming units per gram (cfu/g) in the soil, indicating that aerobic microorganisms capable of metabolizing the fuel oil were present in Area I and Area II.

The carbon dioxide present in Area III has a $\delta^{13}C$, which is heavier than that expected from the biodegradation of the No. 2 fuel oil alone. It is hypothesized that the carbon dioxide in Area III results from either the microbial oxidation of natural organics in the subsurface below the area or a mixing of carbon dioxide from the landfill and the biodegradation of the No. 2 fuel oil. Due to the proximity of Area III to the landfill, it is reasonable to expect that carbon dioxide from the landfill could also be present. Assuming the latter to be the case and that the carbon dioxide from the landfill has a $\delta^{13}C$ value ranging from +10.3 ‰ to +20.0 ‰ (the range in the $\delta^{13}C$ of carbon dioxide reported to have been observed at other landfills), the percent of the total carbon dioxide observed could be estimated from Equation 4. Using Equation 4 and allowing for a fractionation of +3 per mil of the No. 2 fuel oil carbon, it can be calculated that approximately 81% to 85% of the observed carbon dioxide resulted from the biodegradation of the No. 2 fuel oil in Area III.

Although some uncertainty may exist as to whether the carbon dioxide present at the site resulted from the oxidation of the petroleum hydrocarbon/natural organics or emanated from the landfill, stable carbon isotope analysis successfully demonstrated that the carbon dioxide did not result exclusively from the oxidation of the methane present at the site prior to bioventing. In addition, stable carbon isotope analysis allowed adjustment of the amount of carbon dioxide generated from the oxidation of the petroleum hydrocarbons in Area III for carbon dioxide believed to be emanating from the landfill. Because a suitable background bioventing site was not available, this adjustment would have been indeterminable without the use of carbon isotope analysis. The total cost for this evaluation was less than $1,000.

CASE STUDY #2

Site Description

The site, a former underground storage tank (UST) facility, is located in central Oklahoma. When the USTs were removed in 1990, it was observed that petroleum hydrocarbons had leaked into the subsurface. Subsequent site assessment activities defined a separate-phase and dissolved-phase petroleum hydrocarbon plume apparently resulting from the UST release from the site. These site assessment activities also located a previously unknown and unrelated petroleum hydrocarbon plume located to the southeast.

Vacuum extraction and groundwater pumping systems installed at the site have been operational since July 1994. The vacuum extraction system was installed prior to construction of a 90,000 ft^2 warehouse that covers the northern half of the hydrocarbon plume, as shown on Figure 2. The vacuum extraction system is located under the foundation of the new warehouse and was designed to remove the petroleum hydrocarbons present in the vadose zone soils and assist with the recovery of the separate-phase hydrocarbon plume. The groundwater pumping system was installed to halt further migration of and recover the separate-phase hydrocarbon plume and to assist with the recovery of the northern two-thirds of the dissolved-phase plume.

Neither of the remedial systems installed at the site was designed to address the southern one-third of the dissolved hydrocarbon plume. Several technologies were considered for addressing this portion of the plume. Due to the apparent permeability of the saturated soils at the site and the expected biodegradability of the petroleum hydrocarbons, bioremediation (biosparging or bioattenuation) was considered a potential alternative to pump and treat. To determine if bioremediation would be technically feasible at the site, soil samples were collected. The samples were analyzed for several parameters indicative of the subsurface microbiological environment.

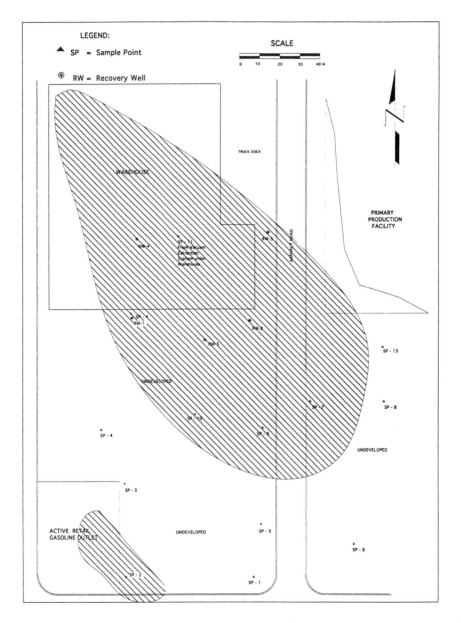

FIGURE 2. Hydrocarbon plume in Case Study #2.

Methodology

Biochemical characterization of the site subsurface microbiological environment consisted of collecting information from both the saturated and unsaturated zones. Soil and groundwater samples were collected and analyzed for total heterotrophic bacteria, contaminant-specific heterotrophic bacteria, total

Kjeldahl nitrogen (TKN), ammonia, and phosphate. In addition, dissolved oxygen in the groundwater and free oxygen in the vadose zone were measured and analyzed in the field from all existing monitoring wells prior to sampling.

All the biochemical parameters measured provided indirect evidence that biodegradation of the petroleum hydrocarbons was occurring naturally. The $\delta^{13}C$ of the carbon dioxide present in the vadose zone by stable carbon isotope analysis was surveyed to provide direct evidence. The survey consisted of collecting 10 soil gas samples (SP-1 through SP-10) from within and in the vicinity of the southern portion of the dissolved-phase plume. One gas sample (SP-11) was collected from the vacuum extraction system during a short (12-h) test of the system. One petroleum hydrocarbon product sample (SP-12) from RW-1 and one background soil sample (SP-13) were collected. Due to site conditions, a product sample from RW-4 could not be obtained.

Vapor samples were collected with a hand-operated vacuum pump that discharged the gas sample into a 1-L Tedlar™ film bag with polypropylene fittings. The product sample was collected with a Teflon™ bailer and placed with no headspace into a 40-mL glass vial with a Teflon™ septum. The soil sample was collected and placed into a glass container.

Based on the literature previously cited, the petroleum hydrocarbon sample could be expected to have a $\delta^{13}C$ value from –34 ‰ to –20 ‰. Because the site is located in an area that is mostly forested, with intermittent prairie grass, the $\delta^{13}C$ value of the natural organics in the area could be expected to range between –29 ‰ to –12 ‰. In addition, it could be expected that carbon dioxide resulting from the oxidation of either of these sources (i.e., petroleum hydrocarbon or natural organics) would have a $\delta^{13}C$ value within approximately 3 ‰ heavier than the substrate.

Results and Discussion

The biochemical characterization of the site samples indicated that the subsurface environment supported the growth of an aerobic, heterotrophic microbial population. The results of the microbial and nutrient analyses suggested the following:

1. Viable bacteria were present in the soil and groundwater (i.e., greater than 10^4 cfu/g).
2. Total heterotrophic bacteria (contaminant-specific) plate counts were from one to two orders of magnitude higher in the downgradient wells compared to the upgradient wells.
3. Soil and groundwater nutrient analysis indicated that nutrients were present in the subsurface in bioavailable forms.
4. Nutrient levels in the groundwater were higher in the background well than in the downgradient wells.

In general, areas with lower dissolved oxygen concentrations corresponded with areas identified as having dissolved hydrocarbon in the groundwater.

The dissolved oxygen concentrations in the groundwater ranged from 0.63 mg/L to 5.64 mg/L. Water saturated with atmospheric oxygen would be expected to have up to approximately 9.0 mg/L dissolved oxygen at standard temperature and pressure. The dissolved oxygen concentration for groundwater in the area based on the collected readings was >3.5 mg/L for groundwater in a background well located on the north side of the site, and between 1.1 mg/L and 2.5 mg/L for groundwater on the downgradient side of the site. The decline in dissolved oxygen concentrations in groundwater from >3.5 mg/L upgradient of the site to <2.5 mg/L downgradient of the site was attributed to uptake by either aerobic respiration of microbes or geochemical factors.

Oxygen levels, measured in the vadose zone soil gas in the lower one-third of the site, ranged from 11.0% to 17.0%, indicating an aerobic environment. Carbon dioxide levels ranged from 5.0% to 15.0%. A strong pattern was evident in the distribution of carbon dioxide, with higher carbon dioxide levels corresponding to areas identified as having dissolved hydrocarbon in groundwater. The vadose zone sample from an area unaffected by xenobiotic compounds had oxygen and carbon dioxide levels of approximately 17% and 5.0%, respectively. The difference between the background levels and atmospheric levels (21.0% for oxygen and <1.0% for carbon dioxide) is likely due to the aerobic oxidation of natural organics present in the shallow soils, and root respiration that uses oxygen and generates carbon dioxide. The samples collected from the vacuum extraction system during a 12-h test were significantly depleted in oxygen (<2%) and enriched in carbon dioxide (>25%), indicating a hypoxic environment under the warehouse. It was hypothesized that the newly constructed warehouse was restricting exchange of atmospheric oxygen with vadose zone soils and that the oxygen present under the warehouse had been used by aerobic microorganisms.

Stable carbon isotope analysis by GIRMS was employed across the southern portion of the dissolved-phase plume to verify that the carbon dioxide observed during the biochemical characterization had resulted from the aerobic oxidation of the petroleum hydrocarbons dissolved in the groundwater. It can be observed from Table 2 that the $\delta^{13}C$ of the background natural organics is approximately –21.0 ‰, whereas the petroleum hydrocarbon present at the site has a $\delta^{13}C$ of approximately –29.0 ‰. The saturate fraction of the hydrocarbon has a $\delta^{13}C$ of –28.9 ‰, whereas the aromatic fraction has a $\delta^{13}C$ of –28.4 ‰. These $\delta^{13}C$ values for the natural organics and petroleum hydrocarbons are well within the ranges expected.

The carbon dioxide $\delta^{13}C$ value in the vicinity of Sample Point (SP) #1, an area unaffected by xenobiotic compounds, was –21.0 ‰, which is identical to that observed for natural organics, indicating little or no fractionation. Based on the $\delta^{13}C$ of carbon substrates and knowledge of the location of the dissolved-phase plumes in the area, the carbon dioxide from SP-3, SP-4, SP-5, SP-8, SP-9, and SP-10 likely results from a mixture of carbon dioxide from the oxidation of natural organics and carbon dioxide that has diffused through the subsurface from the vicinity of the hydrocarbon plumes. The carbon dioxide

TABLE 2. Results of carbon dioxide concentration readings and stable carbon isotope analysis, with all $\delta^{13}C$ reported relative to the PDB standard.

Carbon Dioxide Samples		
Sample No.—Location	Carbon Dioxide, %	$\delta^{13}C$ of Carbon Dioxide, ‰
Sample Point #1	5.5	−21.0
Sample Point #2	15.0	−27.8
Sample Point #3	6.0	−23.6
Sample Point #4	7.5	−26.1
Sample Point #5	5.0	−22.6
Sample Point #6	7.5	−29.1
Sample Point #7	10.5	−29.8
Sample Point #8l	6.0	−23.5
Sample Point #9	5.0	−24.5
Sample Point #10	6.5	−26.7
Sample Point #11	+25.0	−29.5

Gasoline and Natural Organic Carbon Samples		
Sample No.—Location	Carbon Source	$\delta^{13}C$ of Carbon Source, ‰
Sample Point #12	Saturate Fraction (petroleum)	−28.9
Sample Point #12	Aromatic Fraction (petroleum)	−28.4
Sample Point #12	Total (petroleum)	−29.0
Sample Point #12 (duplicate)	Total (petroleum)	−28.9
Sample Point #13	Natural Organics (soil)	−21.0
Sample Point #13 (duplicate)	Natural Organics (soil)	−20.6

samples from SP-6, SP-7, and SP-11, which are located within the hydrocarbon plume area, have $\delta^{13}C$ values ranging from −29.8 ‰ to −29.1 ‰, which is slightly lighter than that observed for the petroleum hydrocarbon.

Based on the stable carbon isotope analysis performed, it can be concluded that the petroleum hydrocarbon plume is being actively degraded by the indigenous microorganisms. This analysis provided verification of biodegradation for a cost of less than $750 versus biochemical analyses, which provided an indirect indication of biodegradation at significantly higher cost.

CONCLUSIONS

The potential value of the stable carbon isotope method in supplementing other biochemical analyses for validating in situ bioremediation or bioattenuation is great. This method can provide several benefits such as:

1. Confirmation of biodegradation: Stable carbon isotope analysis can provide evidence of complete mineralization of the targeted substrate.
2. Reduced analytical costs: Stable carbon isotope analysis is relatively inexpensive ($30 to $65 per sample) compared to other common analyses.
3. Fewer analyses: Fewer samples will be required to provide coverage across a site. This is due to less heterogeneity in gas phase samples compared to sludge, soil, and water samples.
4. Reduction of sampling costs: Fewer samples and the simpler collection procedures for gas samples reduce the amount of time necessary to collect samples.

This monitoring tool can be implemented whenever a definitive demonstration of biodegradation is sought. Biodegradation could be the result of intrinsic activity, biostimulation, or bioaugmentation. One of the obvious uses of the method would be monitoring bioventing and biosparging processes. Another significant application would be in identifying sites where intrinsic biodegradation is occurring. Linking the subsurface gas-phase carbon dioxide to the contaminant(s) present at the site could allow project managers and regulators verification of bioactivity. Combining the reduction in the number of required samples and cost of the stable carbon isotope analysis results in the potential for significant overall savings in monitoring costs at any applicable site.

REFERENCES

Aggarwal, P.K. and R.E. Hinchee. 1991. "Monitoring In Situ Biodegradation of Hydrocarbons by Using Stable Carbon Isotopes." *Environ. Sci. Technol. 25(6):* 1178-1180.
Baedecker, M.J. and W. Back. 1979. "Hydrogeological Processes and Chemical Reactions at a Landfill." *Groundwater 17(5):* 429-437.
Barker, J.F. and P. Fritz. 1981. "Carbon Isotope Fractionation during Microbial Methane Oxidation." *Nature 293:* 289-291.
Chapelle, F.H. 1993. *Ground-Water Microbiology and Geochemistry,* pp. 85-87. John Wiley & Sons, Inc., New York, NY.
Cloud, P.E., I. Friedman, F.D. Sisler, and V.H. Dibeler. 1958. "Microbial Fractionation of Hydrogen Isotopes." *Science 127:* 1394-1395.
Coleman, D.D., W.F. Meents, C. Liu, and R.A. Keogh. 1977. *Isotopic Identification of Leakage Gas from Underground Storage Reservoirs,* Illinois Petroleum Report 111, Illinois State Geological Survey, Urbana, IL.
Coleman, D.D., J.B. Risatti, and M. Schoell. 1981. "Fractionation of Carbon and Hydrogen Isotopes by Methane-Oxidizing Bacteria." *Geochim. Cosmochim. Acta 45:* 1033-1037.
Coleman, D.C., and B. Fry. 1991. *Carbon Isotope Techniques,* p. 174. Academic Press Inc., San Diego, CA.
Craig, H. 1957. "Isotopic Standards for Carbon and Oxygen and Correction Factors for Mass Spectrometric Analysis of Carbon Dioxide." *Geochim. Cosmochim. Acta 12(N1/2):* 133.
Fry, B., W.L. Jeng, R.S. Scalan, P.L. Parker, and J. Baccus. 1978. "δ^{13}C Food Web Analysis of a Texas Sand Community." *Geochim. Cosmochim. Acta 42:* 1299-1302.

Fuex, A.N. 1977. "The Use of Stable Carbon Isotopes in Hydrocarbon Exploration." *J. Geochem. Explor* 6: 139-162.

Galimov, E.M. 1985. *The Biological Fractionation of Isotopes,* Academic Press.

Games, L.M. and J.M. Hayes. 1976. "On the Mechanisms of CO_2 and CH_4 Production in Natural Anaerobic Environments." *Environmental Biogeochemistry,* pp. 51-73. Ann Arbor Science, Ann Arbor, MI.

McMahon, P.B., D.F. Williams, and J.T. Morris. 1990. "Production and Carbon Isotopic Composition of Bacterial CO_2 in Deep Coastal Plain Sediments of South Carolina." *Groundwater* 28(5): 693-702.

Parker, P.L. and J.A. Calder. 1970. "Stable Carbon Isotope Ratio Variations in Biological Systems." In D.W. Hood (Ed.), *Organic Matter in Natural Waters,* pp. 107-127. Institute of Marine Science, U. of Alaska, Pub. No. 1.

Rosenfeld, W.D. and S.R. Silverman. 1959. "Carbon Isotope Fractionation in Bacterial Production of Methane." *Science 130:* 1658-1659.

Schimel, D.S. 1993. *Theory and Application of Tracers,* Academic Press, Inc., San Diego, CA.

Schoell, M. 1980. "The Hydrogen and Carbon Isotopic Composition of Methane from Natural Gases of Various Origins." *Geochimica et Cosmochimica Acta 44:* 649-661.

Suchomel, K.H., D.K. Kreamer, and A. Long. 1990. "Production and Transport of Carbon Dioxide in a Contaminated Vadose Zone: A Stable and Radioactive Carbon Isotope Study." *Environ. Sci. Technol.* 24(12): 1824-1831.

Whiticar, E., and E. Faber. 1985. "Methane Oxidation in Marine and Limnic Sediments and Water Columns." *12th Intl. Mtg. Org. Geochim.*

Whiticar, E., E. Faber, and M. Schoell. 1986. "Biogenic Methane Formation in Marine and Freshwater Environments: CO_2 Reduction *vs.* Acetate Fermentation—Isotope Evidence." *Geochimica et Cosmochimica Acta 50:* 693-709.

Zyakun, A.M., V.A. Bondar, B.B. Namsarayev, and A.I. Nesterov. 1983. "Use of Carbon-Isotope Analysis in Determining the Rates of Microbial Oxidation of Methane in Natural Ecosystems." Translated from *Geokhimiya 5:* 759-765.

AUTHOR LIST

Ahrné, Siv
Lund University
Lab of Food Hygiene & Technology
P.O. Box 124
S-22100 Lund
SWEDEN

Andrilenas, Jeffrey S.
AGRA Earth & Environmental, Inc.
3232 West Virginia Avenue
Phoenix, AZ 85009-1502 USA

Arulgnanendran, Vethanayagam R. J.
New Mexico State University
Dept. of Civil, Agricultural,
    and Geological Engrg.
Box 30001, Dept 3CE
Las Cruces, NM 88003-0001 USA

Barber, Christopher
CSIRO
Division of Water Resources
Private Bag PO
Wembley, Western Australia 6014
AUSTRALIA

Bare, Richard E.
Exxon Research & Engineering
Clinton Township
Route 22 East
Annandale, NJ 08801 USA

Bech, Cathe
SINTEF Applied Chemistry
Environmental Technology
Brattora Research Centre
N-7034 Trondheim
NORWAY

Bennett, Marlene L.
BHP Research
Melbourne Research Labs
245 Wellington Road
Mulgrave, Victoria 3170
AUSTRALIA

Benson, Alvin K.
Brigham Young University
Department of Geology
P.O. Box 24642
Provo, UT 84602-4642 USA

Brockman, Fred J.
Battelle Pacific Northwest
Environmental Sciences Dept.
P.O. Box 999, K4-06
Richland, WA 99352 USA

Burgos, William
VA Polytechnic & State University
Dept. of Civil Engineering
320 Norris Hall
Blacksburg, VA 24061-0246 USA

Calcavecchio, Peter
Exxon Research & Engineering
Clinton Township
Route 22 East
Annandale, NJ 08801 USA

Carpels, Mark
VITO/Flemish Inst. for Technological
    Research
Environment Technology
Boeretang 200
B-2400 Mol
BELGIUM

Cifuentes, Luis A.
Texas A&M University
Department of Oceanography
College Station, TX 77843-3146 USA

Coffin, Richard B.
U.S. Environ. Protection Agency
Environmental Research Lab
1 Sabine Island Drive
Gulf Breeze, FL 32561 USA

Crawford, Ronald L.
University of Idaho
Dept. of Microbiology, Molecular
    Biology, and Biochemistry
Food Research Center 103
Moscow, ID 83844-1052 USA

Davis, Gregory B.
CSIRO, Division of Water Resources
Perth Laboratory
Private Bag PO
Wembley, Western Australia 6014
AUSTRALIA

Diels, Ludo
SCK-CEN/VITO/Flemish
Lab of Genetics and Biotechnology
Boeretang 200
B-2400 Mol
BELGIUM

Dijkhuis, Edwin J.
Bioclear Environ. Biotechnology
P.O. Box 2262
NL-9704 CG Groningen
THE NETHERLANDS

Dott, Wolfgang
Inst. for Hygiene and Environ.
    Medicine
Rwth Aachen, Pauwelsstr. 30
D-52057 Aachen
GERMANY

Douglas, Gregory S.
Arthur D. Little, Inc.
25 Acorn Park
Cambridge, MA 02140-2390 USA

Drake, Evelyn N.
Exxon Research & Engineering
Environmental Science Section
Clinton Township
Route 22 East
Annandale, NJ 08801 USA

Dunbavan, Michael
BHP Engineering
Environmental Technologies Intl.
P.O. Box 3379
Honolulu, HI 96842 USA

Elslander, Helmut
VITO/Flemish Inst. for Technological
    Research
Environment Technology
Boeretang 200
B-2400 Mol
BELGIUM

Faksness, Liv-Guri
SINTEF Applied Chemistry
Environmental Technology
Brattora Research Centre
N-7034 Trondheim
NORWAY

Favre, Alain
ETH Eldgenossiche Technische
    Hochschule Zurich
Biocompatible Matls. Sci. & Engrg.
Wagistrasse 6
CH-8952 Schlieren
SWITZERLAND

Fu, Chunsheng
University of Cincinnati
Dept. of Chemical Engineering
620 Rhodes Hall
Cincinnati, OH 45221 USA

Gao, Chao
University of Cincinnati
Dept. of Chemical Engineering
620 Rhodes Hall
Cincinnati, OH 45221 USA

Govind, Rakesh
University of Cincinnati
Dept. of Chemical Engineering
620 Rhodes Hall
Cincinnati, OH 45221-0171 USA

Gray, Nancy R.
American Chemical Society
1155 16th Street, NW
Washington, DC 20036 USA

Guerin, Louise J.
Innovation Assessment and Research
The World Trade Centre
P.O. Box 462
Melbourne, Victoria 3005
AUSTRALIA

Guerin, Turlough F.
Minenco Bioremediation Services
1 Research Avenue
Bundoora, Victoria 3083
AUSTRALIA

Henssen, Maurice J.C.
Bioclear Environ. Biotechnology
P.O. Box 2262
NL-9704 CG Groningen
THE NETHERLANDS

Hooybergs, Liliane
VITO/Flemish Inst. for Technological
    Research
Environment Technology
Boeretang 200
B-2400 Mol
BELGIUM

Huesemann, Michael H.
Battelle Pacific Northwest
P.O. Box 999, MS P7-41
Richland, WA 99352 USA

Johansen, Bror
SINTEF Applied Chemistry
Environmental Technology
Brattora Research Centre
N-7034 Trondheim
NORWAY

Johnston, Colin D.
CSIRO, Division of Water Resources
Perth Laboratory
Private Bag PO
Wembley, Western Australia 6014
AUSTRALIA

Kerr, Jill M.
Exxon Production Research Co.
P.O. Box 2189
Houston, TX 77252-2189 USA

Keuning, Sytze
Bioclear Environ. Biotechnology
P.O. Box 2262
NL-9704 CG Groningen
THE NETHERLANDS

Kinnaer, Luc
VITO/Flemish Inst. for Technological
    Research
Environment Technology
Boeretang 200
B-2400 Mol
BELGIUM

Korr, Josef
German Ministry of Defense
Fontainengraben 150
53123 Bonn
GERMANY

Li, Dong X.
Unocal Corporation
Environmental Tech. Group
376 South Valencia Avenue
Brea, CA 92621 USA

MacPherson, Jr., James R.
AGRA Earth & Environmental, Inc.
7477 SW Tech Center Drive
Portland, OR 97223-8025 USA

Mammel, Jr., William
International Lubricants, Inc.
7930 Occidental South
Seattle, WA 98108 USA

Mariner, Paul E.
INTERA, Inc.
6850 Austin Center Blvd., Suite 300
Austin, TX 78731 USA

Marley, Michael C.
Envirogen, Inc.
480 Neponset Street
Canton, MA 02021 USA

Maue, Georg
Technical University Berlin
Fachgebiet Hygiene
Amrumer Str. 32
D-13353 Berlin
GERMANY

Mayer, Jörg
ETH Eldgenossiche Technische
    Hochschule Zürich
Biocompatible Matls. Sci. & Engrg.
Wagistrasse 6
CH-8952 Schlieren
SWITZERLAND

McDonald, Thomas J.
Texas A&M University
Geochem. & Environ. Research
    Group
833 Graham Road
College Station, TX 77845 USA

McKinney, Daene C.
The University of Texas at Austin
Dept. of Petroleum & Systems Eng.
Austin, TX 78712 USA

McMillen, Sara J.
Exxon Production Research Co.
P.O. Box 2189
Houston, TX 77252-2189 USA

Moore, Roy E.
AGRA Earth & Environmental, Inc.
7477 SW Tech Center Drive
Portland, OR 97223-8025 USA

Mueller, James G.
SBP Technologies Inc.
1 Sabine Island Drive
Gulf Breeze, FL 32561-3999 USA

Nirmalakhandan, Nagamany
New Mexico State University
Dept. of Civil, Agricultural,
    and Geological Engineering
Box 30001, Dept 3CE
Las Cruces, NM 88003-0001 USA

Novak, John T.
VA Polytechnic & State University
Dept. of Civil Engineering
320 Norris Hall
Blacksburg, VA 24061-0246 USA

Patterson, Bradley M.
CSIRO, Division of Water Resources
Private Bag PO
Wembley, Western Australia 6014
AUSTRALIA

Payne, Kelly L.
Brigham Young University
Department of Geology
P.O. Box 24642
Provo, UT 84602-4642 USA

Persson, Anders
ANOX AB
Ideon Research Park
S-22370 Lund
SWEDEN

Petitmermet, Marc
ETH Eldgenossiche Technische
    Hochschule Zürich
Biocompatible Matls. Sci. & Engrg.
Wagistrasse 6
CH-8952 Schlieren
SWITZERLAND

Pfanstiel, Steven
University of Cincinnati
Dept. of Chemical Engineering
Cincinnati, OH 45221 USA

Pfiffner, Susan M.
Oak Ridge National Laboratory
Environmental Sciences Division
Oak Ridge, TN 37831-6038 USA

Pinkart, Holly C.
Oak Ridge National Laboratory
c/o University of Tennessee
Center for Environ. Biotechnology
10515 Research Drive, Suite 300
Knoxville, TN 37932-2575 USA

Pope, Gary A.
The University of Texas at Austin
Dept. of Petrol. & Geosystems Eng.
Austin, TX 78712 USA

Prince, Roger C.
Exxon Research & Engineering
Clinton Township
Route 22 East
Annandale, NJ 08801 USA

Pruess, Alan J.
AGRA Earth & Environmental, Inc.
7477 SW Tech Center Drive
Portland, OR 97223-8025 USA

Quednau, Mikael
Lund University
Lab of Food Hygiene & Technology
P.O. Box 124
S-22100 Lund
SWEDEN

Ramstad, Svein
SINTEF
Applied Chemistry
Environmental Technology Group
N-7034 Trondheim
NORWAY

Requejo, Adolpho G.
Texas A&M University
Geochem. & Environ. Research
    Group
833 Graham Road
College Station, TX 77845 USA

Ringelberg, David B.
University of Tennessee
Center for Environ. Biotechnology
10515 Research Drive, Suite 300
Knoxville, TN 37932-2575 USA

Rösler, Ursula
PLUS Endoprothetik AG
Erlenstrasse 4b
CH-6343 Rotkreuz
SWITZERLAND

Rothenburger, Stephen J.
Exxon Research & Engineering
Clinton Township
Route 22 East
Annandale, NJ 08801 USA

Rouse, Bruce A.
The University of Texas at Austin
Dept. of Petrol. & Geosystems Eng.
Austin, TX 78712-1076 USA

Saber, Diane L.
Saber Environ. Consultants, Inc.
21155 Amberly Drive
Kildeer, IL 60047 USA

Saberiyan, Amy G.
AGRA Earth & Environmental, Inc.
7477 SW Tech Center Drive
Portland, OR 97223-8025 USA

Schröder, Wolfgang
Head Dept. of Finance of Lower
    Saxony
Waterloostrasse 4
30169 Hannover
GERMANY

Schuman, David
VA Polytechnic & State University
Dept. of Civil Engineering
320 Norris Hall
Blacksburg, VA 24061-0246 USA

Shah, Brigitte
ETH Eldgenossiche Technische
    Hochschule Zürich
Biocompatible Matls. Sci. & Engrg.
Wagistrasse 6
CH-8952 Schlieren
SWITZERLAND

Shanahan, Alka
International Lubricants Inc.
7930 Occidental South
Seattle, WA 98108 USA

Sheehy, Alan J.
University of Canberra
Faculty of Applied Science
Kirinari Street, Building 3
Bruce, ACT 2617
AUSTRALIA

Stair, Julia O.
University of Tennessee
Center for Environ. Biotechnology
10515 Research Drive, Suite 300
Knoxville, TN 37932-2575 USA

Stokley, Karen E.
Exxon Research & Engineering
Clinton Township
Route 22 East
Annandale, NJ 08801 USA

Stubben, Melissa A.
University of Arizona
3455 North Edith Blvd. #D
Tucson, AZ 85716 USA

Studer, James
INTERA, Inc.
1650 University Blvd. NE, Suite 300
Albuquerque, NM 87102-1732 USA

Sutton, Susan D.
University of Tennessee
Center for Environ. Biotechnology
10515 Research Drive, Suite 300
Knoxville, TN 37932 USA

Sveum, Per
SINTEF Applied Chemistry
Environmental Technology
Brattora Research Centre
N-7034 Trondheim
NORWAY

Tabak, Henry H.
U.S. Environ. Protection Agency
Biosystems Development Section
26 W. Martin Luther King Jr. Drive
Cincinnati, OH 45268 USA

Tiedje, James M.
Michigan State University
Center for Microbial Ecology
540 Plant & Soil Science Bldg.
East Lansing, MI 48824-1325 USA

Trust, Beth A.
National Research Council
c/o U.S. EPA/GBERL
1 Sabine Island Drive
Gulf Breeze, FL 32561 USA

van der Waarde, Jaap J.
Bioclear Environ. Biotechnology
P.O. Box 2262
NL-9704 CG Groningen
THE NETHERLANDS

Van de Velde, Kevin D.
Envirogen, Inc.
Princeton Research Center
4100 Quakerbridge Road
Lawrenceville, NJ 08648 USA

Vanhoutven, Diana
VITO/Flemish Inst. for Technological
    Research
Environment Technology
Boeretang 200
B-2400 Mol
BELGIUM

VanRoy, Sandra
VITO/Flemish Inst. for Technological
    Research
Environment Technology
Boeretang 200
B-2400 Mol
BELGIUM

Wagner, David M.
Envirogen Inc.
4100 Quakerbridge Road
Lawrenceville, NJ 08646-4702 USA

Weth, Dieter
Prof. Mull & Partner GmbH
Osteriede 5
30827 Garbsen
GERMANY

White, David C.
University of Tennessee
Center for Environ. Biotechnology
10515 Research Drive, Suite 300
Knoxville, TN 37932-2575 USA

Whitley, Jr., G. Allen
The University of Texas at Austin
Dept. of Petrol. & Geosystems Eng.
Austin, TX 78712-1076 USA

Wintermantel, Erich
ETH Eldgenossiche Technische
    Hochschule Zürich
Biocompatible Matls. Sci. & Engrg.
Wagistrasse 6
CH-8952 Schlieren
SWITZERLAND

Yan, Xuesheng
University of Cincinnati
Dept. of Chemical Engineering
Cincinnati, OH 45221 USA

Zhou, Enning
University of Idaho
Dept. of Microbiology, Molecular
    Biology, and Biochemistry
Food Research Center 103
Moscow, ID 83844-1052 USA

Zhou, Jizhong
Michigan State University
Center for Microbial Ecology
540 Plant & Soil Science Bldg.
East Lansing, MI 48824-1325 USA

INDEX

acenaphthene, 127, 129-132
acetate, 43, 244, 257
acetone, 81, 167
Acinetobacter, 142
actinomycete(s), 52, 56
activated carbon, 149
activated sludge, 12, 99, 174
adaptation, 127
adsorption, 20, 27-28, 203, 205-207, 216, 220
aeration, 42, 63, 141, 193-194, 198-199, 248
Ag, *see* silver
agar diffusion overlay test, 225, 227, 229, 233-234
aged hydrocarbon, 2
air-filled porosity, 194-197, 200
air injection, 43, 115, 119-124, 201
air sparging, 63, 201
Alaska, 257
Alcaligenes, 67, 69, 129
algae, 56
aliphatic hydrocarbons, 74
alkanes, 2, 3, 7, 11, 15, 45, 56, 212
American Society for Testing and Materials (ASTM), 174, 225-226, 229, 231, 233
ammonia (NH_3), 21, 43-44, 81, 253
ammonia monooxygenase (AMO), 43
ammonium nitrate, 12
AMO, *see* ammonia monooxygenase
anaerobic, 45, 47, 52, 67, 74, 75, 84, 89, 248, 257
anaerobic bacteria, 45
anthracene, 127, 129, 131, 132
apparent resistivity, 105, 108
aqueous, 103, 166, 203, 208
Arizona, 110, 112
aromatic compounds, 18, 19, 66
aromatic fraction(s), 3, 12, 13, 254-255
aromatic hydrocarbon(s), 9, 13, 14, 17, 27-28, 41, 74, 132, 145, 233
aromatics, 2-4, 6, 11, 13, 37, 169
ash, 149, 223-224, 227-234
asphaltenes, 2

ASTM, *see* American Society for Testing and Materials
attenuation rate, 191
augmentation, 161
Australia, 176, 183, 194
Austria, 225, 227
Azoarcus, 67, 69, 75

B. cepacia, *see* Burkholderia cepacia
B. pickettii, *see* Burkholderia pickettii
BAC, *see* biological activated carbon
bacteria, 9, 20-21, 41, 43, 45-47, 52, 56-57, 74, 75, 78, 81, 88, 91, 127-129, 132, 133, 148-149, 153-155, 174, 185-187, 233-235, 237-239, 243, 252-253, 256
bacterial isolate(s), 45, 74, 130, 131, 133
bacterial population, 49, 128, 185
bacterium, 41-42, 74, 75, 132, 233
beach models, 87, 88
Belemnitella, 243
Belgium, 151
bench-scale reactors, 141, 146
benzene, 42, 67, 106, 108
benzene, toluene, ethylbenzene, and xylenes (BTEX), 42, 43, 67, 106, 108
bioattenuation, 241, 251, 255
bioaugmentation, 176, 256
bioavailability, 27, 59, 176, 178, 209
biodegradability, 3, 7, 17, 20, 27-28, 59, 97-100, 102, 209, 248, 251
biodegradation kinetics, 203, 208-209
biodegradation rate, 124, 186, 204
biofeasibility, 185-186
biofiltration, 124
biogeochemistry, 257
biokinetic, 203-204, 208-209
bioleaching, 223
biological activated carbon (BAC), 77, 81-84
biomarker(s), 2, 3,5, 11, 19-20, 27, 28, 44-47, 49-50, 53, 57
BIOPLUME, 162
bioslurry, 161

biosparging, 241, 251, 256
biostimulation, 46-47, 176, 256
biotreatability study, 185-186, 191
bioventing, 43, 63, 115-117, 123, 124,
    126, 241, 248, 250-251, 256
bioventing monitoring, 115, 116
biphasic, 31, 35
biphenyl, 46
bromide, 68, 151
BTEX, see benzene, toluene,
    ethylbenzene, and xylenes
bulking agents, 2
Burkholderia, 65, 67, 69
Burkholderia cepacia, 65-71
Burkholderia pickettii, 67, 69, 71

cadmium (Cd), 231
calcite, 242-243
carbon-12, 241-242
carbon-13, 241-243, 245
carbon dioxide (CO$_2$), 59-63, 93-95,
    98-100, 103, 115, 117, 142, 149, 152,
    185-186, 188, 190-191, 193, 201, 205,
    208-209, 225, 233-239, 241-246, 248-
    251, 253-257
carbon dioxide production, 152, 193
casamino acid, 99
catechol, 42-43
catechol 2,3-dioxygenase, 43
cation exchange capacity (CEC), 205
Cd, see cadmium
CEC, see cation exchange capacity
cell culture techniques, 223-224, 231, 234
CFU(s), see colony-forming unit
CH$_4$, see methane
characterization, 11-13, 29, 31, 39, 45-46,
    52, 57, 74, 132, 141, 153, 162,
    185-186, 211-212, 219, 221, 252-254
chloroform, 235
chromium (Cr), 159-160
chrysene, 19, 21
clay, 21, 30, 106
CO$_2$, see carbon dioxide
coal tar, 128
coastal, 85, 96, 257
co-contamination, 145
CO$_2$ evolution, 60, 61
cognitive style(s), 175, 178-179, 180-183
colony-forming units (CFU), 41, 129,
    130, 142, 144, 149, 152-153, 186,
    234, 237, 250, 253

Colorado, 17
columns, 29-31, 34, 36-37, 45, 77-82, 84,
    85, 87, 88, 96, 154, 214, 216, 257
Comamonas, 129
cometabolic, 46, 178
cometabolism, 239
comparative, 59
complete degradation, 35, 37
compost, 147-154
composting, 147-148, 176
Connecticut, 68
consortium, 57, 143
continuous-flow column(s), 78, 79, 80,
    88
continuous-flow mesoscale basins, 86,
    87
continuous monitoring, 115, 124
controlled release, 176
copper (Cu), 231, 235
cost estimates, 138
Cr, see chromium
creosote, 159-160
crude oil, 1, 7-9, 11, 12, 15-18, 20, 28, 30,
    77, 78, 81, 85, 93, 96
Cu, see copper
cytotoxicity, 224-225, 227, 229, 231-232,
    234

2,4-D, see 2,4-dichlorophenoxyacetic
    acid
database, 66, 72, 74, 136, 138-140,
    185-186
database solution, 136, 139
d13C values, 233-239, 243-246, 249,
    254-255
degradation efficiency, 143
degradation rate(s), 31, 35, 36, 41, 43-44,
    103, 127, 128, 153, 162, 190-191,
    193-196, 199-200
dehydrogenase, 43-44, 59-63, 225, 227,
    229, 231
denitrifiers, 75
denitrifying, 67, 74
dense, nonaqueous-phase liquids
    (DNAPLs), 212
desorption, 11-13, 203, 205-207
diauxic, 94
diauxy, 94
2,4-dichlorophenoxyacetic acid (2,4-D),
    41, 46
diesel, 17, 116, 119, 128-130, 149, 151,

153-154, 188, 190, 193-194,
199-200
diffusion, 27, 119, 120, 194-195, 199-200,
203, 205, 208, 225-227, 229, 231,
233-234
digester, 93
dispersion, 117
dissolution, 135
dissolved organic carbon (DOC), 233-
237
dissolved oxygen (DO), 16, 40, 44, 59,
63, 77, 124, 159, 199, 204, 253-254
dissolved-phase plume, 251, 253-254
DNAPL(s), *see* dense, nonaqueous-phase
liquids
DO, *see* dissolved oxygen
DOC, *see* dissolved organic carbon
dual-inlet gas isotope ratio mass
spectrometer, 243
dynamic respiration measurement, 124

E. coli, see Escherichia coli
EDTA, *see* ethylenediaminetetraacetic
acid
electrical resistivity, 105, 108, 112
electrode, 80, 108
electronic data processing, 136
enrichment, 117
Escherichia coli, 65, 66, 68, 70-72
esterase, 59, 81-84
ethylbenzene, 42, 67, 106, 108
ethylene, 57, 234
ethylenediaminetetraacetic acid (EDTA),
205
European Union, 175, 183
ex situ bioremediation, 203
experimental procedure, 91, 206
extraction well, 43

fatty acid(s), 44-45, 47, 49-50, 52-53, 56-
57
FDA, *see* fluorescein diacetate
Fe, *see* iron
fermentation, 128, 129, 244, 257
fertilizer(s), 2, 3, 12, 16, 78, 86-88, 96
field desorption mass spectroscopy, 11,
12
field experiments, 77, 85, 87, 88
fish meal, 77, 81-84, 93-95
fluoranthene, 19, 129-132, 233-238

fluorescein diacetate (FDA), 59-63, 81,
85
fly ash, 223-224, 227-234
formalized risk evaluation, 138
fractionation, 209, 238, 241, 243, 245-246,
251, 254, 256-257
France, 209
fuel, *see* crude oil, diesel, fuel oil,
gasoline, jet fuel, JP-5, No. 2 fuel
oil, petroleum, *and* weathered
diesel
fuel oil, 246-251
fungus, fungi, 9, 153, 232

ganglia, 216
gas chromatography (GC), 1-3, 5, 7, 8,
17, 19-20, 22, 29-31, 52, 59, 82, 93,
142, 149, 151-152, 206, 216, 235-237
gas chromatography/mass
spectrometry (GC/MS), 1-3, 5, 17
gas isotope ratio mass spectrometer
(GIRMS), 243, 249, 254
gasoline, 106-109, 112, 116, 126, 142,
159, 188, 190, 249, 255
gasoline vapors, 249
gas tracers, 211-212, 219
GC, *see* gas chromatography
GC/MS, *see* gas chromatography/mass
spectrometry
gene probe(s), 41, 42-44, 46, 47, 56, 66,
68, 73
gene probing, 50
geographic information system (GIS),
140
geophysics, 112, 113
Germany, 98, 128, 133, 135-136, 140,
183, 227, 232
GIRMS, *see* gas isotope ratio mass
spectrometer
GIS, *see* geographic information system
glucose, 129-132, 149, 234-235, 237-238
gravel, 21, 81, 85, 93, 96, 106, 116
groundwater pumping, 251
growth rate, 141, 144-145

H_2, *see* hydrogen
half-life, half-lives, 185-186, 190-191
headspace, 169, 206, 235-236, 253
heavy metal(s), 145, 223
Henry's law, 212

heterogeneity, 1, 20, 84, 220, 256
heterotrophic bacteria, 21, 252-253
heterotrophic respiration, 93-95
hexadecane, 42
n-hexadecane, 149
hexane, 81, 93, 195
Hg, *see* mercury
HgCl₂, *see* mercuric chloride
hierarchy of treatability studies, 157
H₂O₂, *see* hydrogen peroxide
hopane, 2, 3, 7-9, 17, 19-21, 23-28
humic acids, 21
hydraulic oil(s), 98, 103
hydrocarbon characterization, 11, 12
hydrocarbon degradation, 94, 199
hydrocarbon degraders, 149, 153
hydrocarbon-degrading
    microorganisms, 242
hydrocarbon plume, 116-118, 121, 122,
    251-252, 255
hydrocarbon types, 11-16
hydrogen (H₂), 11-13, 16, 30, 99, 188,
    206, 244, 256-257
hydrogen peroxide (H₂O₂), 30
hydrolysis, 59-63, 85

Idaho, 97-99
Illinois, 103, 256
in situ respiration, 115, 118, 126, 201
in situ respirometry, 119, 123, 125
Indiana, 68
indigenous microbes, 97
indigenous microorganisms, 248, 255
induction, 41, 105, 183
inhibition, 167, 169, 172, 174
injection well, 43, 119, 122-124
inoculation, 129
inoculum, 12, 98, 99, 128-130, 142
INSA (German database), 136
insecticides, 174
interval resistivity, 108, 109
intrinsic bioremediation, 43, 233
iron (Fe), 148, 199
isoprenoid(s), 2, 3, 7
isotope, 233, 235-236, 239, 241-243, 244,
    245, 246, 248-251, 253-257

Japan, 201
jet fuel, 30, 46
JP-5, 43, 46

kinetic(s), 43, 84, 85, 141, 185, 187-188,
    191, 203, 208-209
Kwajalein, 49-50, 52-53, 56-57

landfarming, 12, 17
landfill, 248, 250-251, 256
land treatment, 16
Li, *see* lithium
light hydrocarbons, 47
lipid analysis, 50, 57
liquid culture, 98, 128
lithium (Li), 115, 116, 118, 120, 123, 126
loam, 12, 21, 27, 30, 166, 205
lubricant(s), 97-103
lubricating oil, 99, 151

mass balance, 12, 39, 78, 84, 88, 219,
    241-242, 244
mass transfer, 30
Max Bac, 77, 81-84
mercuric chloride (HgCl₂), 99, 149-150,
    206
mercury (Hg), 41, 46-47
mesocosm, 12, 77, 85
methane (CH₄), 40, 43-47, 149, 195, 201,
    235, 243-244, 246, 248-251, 256-257
methane monooxygenase (MMO), 40
methanogenesis, 244
methanol, 2, 43-44
methanol dehydrogenase, 43-44
methanotroph(s), 44-45, 47, 57, 246
methanotrophic, 40, 43-45
methylene chloride, 2, 20-22, 30, 234-236
Methylomonas, 66, 67, 69
methylotroph, 66
Michigan, 17
microaerophilic, 67
microbe, 188
microbial activity, 56, 63, 81, 85, 87, 94,
    95, 99, 152, 199, 201
microbial community structure, 39, 56
microbial population(s), 40-41, 46, 56,
    161, 253
microcosm(s), 9, 18, 20, 30-31, 33, 36, 41-
    42, 45, 74
microflora, 41, 65, 128, 147-148, 174
Microtox®, 174
mineralization, 41-43, 47, 209, 233,
    236-237, 241, 246, 256
mineral oil, 59-63, 144-145

mixed culture(s), 127-129, 131, 132, 176
MMO, *see* methane monooxygenase
Mn, *see* manganese
monitoring well(s), 105-107, 108, 112,
    117, 124-126, 253
Monod equation, 188
Montana, 53, 57
most probable number (mpn), 20, 41-44,
    50, 60
motor oil, 11, 12, 15-17
mpn, *see* most probable number

N, *see* nitrogen
naphthalene, 21, 41-43, 46-47, 131
NAPL(s), *see* nonaqueous-phase liquid
NAPL degradation rate(s), 193-196, 199-
    200
natural attenuation, 39
Netherlands, 112
New Jersey, 66
New Mexico, 166
New York, 9, 47, 74, 113, 126, 167, 174,
    205, 221, 239, 256
NH$_3$, *see* ammonia
Ni, *see* nickel
niche, 72, 181
nickel (Ni), 231, 235
nitrate, 12, 21, 81, 93
nitrification, 174
nitrite, 21, 81
nitrogen (N), 13, 4, 142, 149, 206-207
Nitrosolobus, 66, 67, 69
Nitrosomonas, 66, 67, 69
nitrous oxide, 43
No. 2 fuel oil, 246, 248-251
nonaqueous-phase liquid(s) (NAPL),
    193-197, 199-200, 211-214, 218-220
Norway, 220-221
nucleic acid(s), 39-40, 43, 45-46, 74, 155,
    233, 235-239
nutrient addition, 29, 31-35, 38, 194
nutrient amendment, 49, 61, 62

O$_2$, *see* oxygen
oil and grease, 12
oil recovery, 162, 221
oil spill, 27, 77, 87, 96
oily sludge, 147, 149, 151
Oklahoma, 251
oleanane, 19, 21, 27

open system, 77, 78, 84, 85, 87, 88, 199
optimization, 124, 155
organic amendments, 176
organic matter, 21, 27, 205, 246, 257
organic nutrient, 87, 95
organizational innovation, 178
oxygen (O$_2$), 188, 195, 199
oxygenase, 44
oxygen sensor(s), 115-117, 119, 120, 121,
    123-126, 193-195, 197, 200-201
oxygen uptake, 115, 119, 124, 125, 167,
    204, 206-209, 242
oxygen uptake rate, 167
oxygen utilization, 193, 199

P, *see* phosphorus
P. mendocina, *see* Pseudomonas mendocina
P. putida, *see* Pseudomonas putida
PAH(s), *see* polycyclic aromatic
    hydrocarbon
particles, 28, 91, 216
particle size distribution, 207
partitioning, 211-214, 216-221
PCB(s), *see* polychlorinated biphenyl
PCE, *see* tetrachloroethylene
PCP, *see* pentachlorophenol
PCR, *see* polymerase chain reaction
PDB, *see* Pee Dee belemnite
peat, 60-63, 147, 149-151, 153-154
Pee Dee belemnite (PDB), 235, 243-244,
    250, 255
pentachlorophenol (PCP), 159-160
perfluorocarbon(s), 211, 212, 213,
    219-221
pesticide, 166
petroleum hydrocarbon(s), 2, 11, 12, 13,
    17, 19, 21, 29, 38, 42, 47, 49-50, 96,
    106, 116, 118, 126, 161, 241, 243,
    246-247, 249, 251, 253-255
pH, 3, 12, 20-21, 52, 59, 68, 81, 99, 142,
    149-150, 152, 186, 188, 205, 227,
    231, 234
pharmaceutical, 166
phenanthrene, 21-22, 27, 127, 129-132
phenol, 65, 203-208, 235
phosphate, 12, 20, 43, 149, 234, 253
phospholipid fatty acid, 50, 52
phosphorus (P), 2, 99, 149, 186, 190, 205
phytane, 3
pilot plant, 224
pilot testing, 248

platinum (Pt), 225, 235
plugging, 206
polychlorinated biphenyl(s) (PCB), 41,
    50, 56, 57
polycyclic aromatic hydrocarbon(s)
    (PAHs), 2, 3, 6, 9, 17, 19-21, 27-28,
    41, 46, 127-133, 145, 233
polymerase chain reaction (PCR), 40-41,
    43, 46, 65, 66, 68-75, 148, 151, 155
porous medium (-ia), 105, 126, 213-214
precipitation, 2, 242
primer(s), 40-41, 65-74, 75, 148, 151, 155
pristane, 3, 77, 82, 93, 94, 96
pseudomonads, 47
Pseudomonas, 65, 67, 69, 74, 129-132, 155
Pseudomonas mendocina, 67, 69
Pseudomonas putida, 24, 67, 69, 71
Pt, see platinum
pump-and-treat, 65, 251
pyrene, 19, 21, 129, 130, 132
p-xylene, see xylene

quartz, 235, 249
quinone, 45

radius of influence (ROI), 123, 124
randomly applied polymorphic (RAPD),
    40, 147-148, 153-155
RAPD, see randomly applied
    polymorphic DNA
rapeseed, 97, 100
refinery, 12, 19-20, 22
refinery soil, 19-20, 22
residual military pollution, 135
residual saturation, 214
resin, 66
resistance, 29, 41, 127
respiration measurement, 117-120, 122,
    124
respirometer, 167, 204-205
respirometry, 119-123, 125, 147-148, 152,
    185, 209
respirometry test, 119-123, 125, 185
risk assessment, 136, 138, 162, 174
ROI, see radius of influence
root(s), 242, 254

salicylate, 41-42
salinity, 81
sand, 21, 27, 85, 96, 106, 116, 194, 211,

    214, 216-220, 249, 256
sandy, 12, 30, 60-62, 66, 85, 96, 142, 166,
    194, 201
SARA, see Superfund Amendments and
    Authorization Act
saturated zone, 43, 211
saturates, 2-4, 6, 11, 13, 15-17
Savannah River Integrated
    Demonstration (SRID), 47
scale up, 143
screening test(s), 127, 128, 203
SDS, see sodium dodecyl sulfate
Se, see selenium
selenium (Se), 93, 161, 169
sensitivity, 65, 70-74
shoreline bioremediation, 77, 78, 86-88,
    96
signature lipid biomarker, 49-50
silt, 21, 106, 116, 205
silt loam, 205
silty, 116
silver (Ag), 206
simulation, 79, 85, 91
site characterization, 162
slag, 223-224, 227-231, 233-234
sludge(s), 1, 2, 12, 99, 141, 145, 147,
    149-151, 153-154, 174, 224, 256
slurry bioreactor, 204
slurry bioremediation, 12
slurry-phase, 17
slurry reactor(s), 29-31, 35-36, 203-204,
    207-208
slurry treatment, 16, 160
slurry wall, 162
sodium dodecyl sulfate (SDS), 66
soil column(s), 30-31, 45, 47, 57
soil flushing, 161
soil moisture, 20, 168, 204-205
soil vapor extraction (SVE), 124, 125,
    142, 201
soil venting, 126
soil washing, 161, 178
solid-phase, 17
solvent, 2, 21, 44
sorption, 29, 188
South Carolina, 43, 45, 103, 257
sparging, 63, 201 (see also air sparging,
    biosparging)
spatial variability, 2
Sphingomonas, 142, 233-234
SRID, see Savannah River Integrated
    Demonstration

stabilization, 178
stable carbon isotope(s), 233, 235, 239, 241-242, 245, 246, 248, 250-251, 253-257
stick water, 77, 81-84
Sturm Test, 97, 99, 101, 103
subsurface oxygen sensor, 115-117, 119
sulfate, 20
Superfund Amendments and Authorization Act (SARA), 1
surfactant, 50, 52, 176
SVE, *see* soil vapor extraction
Sweden, 147, 151
Switzerland, 223-224, 231

tar oil, 128-130
Tc, *see* technetium
TCE, *see* trichloroethylene
TDS, *see* total dissolved solids
technetium (Tc), 67
Teflon™, 194, 205, 249, 253
TEP, *see* triethyl phosphate
tetrachloroethylene, 50, 56
Texas, 18, 220-221, 239, 256
Thiobacillus, 49-50, 53, 55-56
TKN, *see* total Kjeldahl nitrogen
toluene, 41-44, 65-67, 74, 75, 106, 108, 248
toluene dioxygenase, 43-44
total dissolved solids (TDS), 105, 108-110, 112
total Kjeldahl nitrogen (TKN), 21, 93, 253
total petroleum hydrocarbons (TPH), 2, 3, 7, 8, 11-13, 19-22, 29-31, 34-38, 42-43, 106, 108, 161
toxicity, 52, 98, 103, 165-167, 169, 171-174, 223, 225, 227, 230-234
TPH, *see* total petroleum hydrocarbons
tracer test(s), 212, 219-221
treatability, 157, 159-160, 163, 185
treatability study(-ies), 157, 159-160, 163
treatment train, 162
trichloroethylene (TCE), 40, 50, 65, 74, 211, 246 40, 43-45, 50, 56, 65, 66, 72, 73, 211, 214, 216-220, 246
triethyl phosphate (TEP), 43
triphenyltetrazolium chloride (TTC), 59, 63
triterpane, 3, 5, 7, 19-20
TTC, *see* triphenyltetrazolium chloride

underground storage tank (UST), 251
United Kingdom, 179, 182
United States, 29, 175-176, 211
unsaturated zone, 201, 211 (*see also* vadose zone)
UST, *see* underground storage tank
Utah, 106, 107, 110, 112, 113
utilization rate(s), 190, 193, 200

vacuum extraction (VE), 138, 251, 253-254
vadose zone, 43, 115, 200-201, 211-212, 219, 248-249, 251, 253-254, 257 (*see also* unsaturated zone)
Valdez, 27, 94, 96
vapor extraction wells, 116, 124-126
VE, *see* vacuum extraction
very low frequency electromagnetic (VLF EM), 105, 110, 111
Vibrio, 57
Virginia, 38
viscosity, 99, 103
vitrification, 223, 234
vitrified waste, 227-228, 233-234
VLF EM, *see* very low frequency electromagnetic
volatile hydrocarbon(s), 248
volatilization, 2, 46-47, 188, 206

wafer, 203-205, 207-209
Washington, DC, 28, 52
water content, 204, 207
weathered diesel, 193-194, 200
weathering, 12, 19, 52
wetlands, 162

xenobiotic, 17, 57, 246, 254
xenobiotic compounds, 254
xylenes, 42-43, 67, 106, 108
p-xylene, 67

yeast extract, 98, 99, 205

zinc (Zn), 231
Zn, *see* zinc